Recent Advances for Coal Energy in the 21st Century

Edited by Yongseung Yun

Published in London, United Kingdom

Recent Advances for Coal Energy in the 21st Century
http://dx.doi.org/10.5772/intechopen.104060
Edited by Yongseung Yun

Contributors
Kalpana C Maheria, Ajay K. Dalai, Henilkumar M. Lankapati, Tomoki Shimanishi, Taku Shimizu, Naoko Shimazaki, Ken Takahashi, Shigeo Nakajima, Anand Pd Sinha, Rahul Rai, Ashok Kumar Asthana, Neha Choudhary, Praveen Chandra Jha, Oluwafikemi Iji, Ahmad Padhil, Nurhayati Rauf, Ayu Reski Ilahi, Ali Can Sivri

First published in London, United Kingdom, 2024 by IntechOpen
IntechOpen is the global imprint of INTECHOPEN LIMITED, registered in England and Wales, registration number: 11086078, 5 Princes Gate Court, London, SW7 2QJ, United Kingdom

British Library Cataloguing-in-Publication Data
A catalogue record for this book is available from the British Library

Additional hard and PDF copies can be obtained from orders@intechopen.com

Recent Advances for Coal Energy in the 21st Century
Edited by Yongseung Yun
p. cm.
Print ISBN 978-1-83768-580-6
Online ISBN 978-1-83768-581-3
eBook (PDF) ISBN 978-1-83768-582-0

For EU product safety concerns:
IN TECH d.o.o., Prolaz Marije Krucifikse Kozulić 3, 51000 Rijeka, Croatia,
info@intechopen.com or visit our website at intechopen.com.

Meet the editor

Dr. Yongseung Yun obtained a BSc from Yonsei University, Korea, in 1979, a master's degree from Korea Advanced Institute of Science and Technology (KAIST), in 1981, and a Ph.D. from the University of Utah, USA, in 1990. He is a chair researcher at the Institute for Advanced Engineering (IAE), Korea. From 1981 to 1984 Dr. Yun researched coal–oil agglomeration and fluidized bed coal combustion at KAST. For his Ph.D., he worked on the low-temperature air oxidation of coal and developed the TG/MS instrument for coal analysis. At Brown University in 1991–1992, Dr. Yum researched pretreatment methods to enhance coal liquefaction yield. Since 1993 he has concentrated on process developments to produce clean synthetic gas from coal, petcoke, and wastes including biomass. Currently, he heads the blue hydrogen project that includes 30 tons/day of petcoke gasification, syngas cleaning, and PSA purification for 2 tons/day of high-purity hydrogen production. During 2013–2018, he was the president of the Korea Association of Waste to Energy (KAWET). He is currently an auditor for the Korea Society of Waste Management and vice president of the Korea DME Association. He has been the chief editor of Korean Industrial Chemistry News for eight years.

Contents

Preface

Coal has been utilized heavily during the last century and will likely continue to be used for at least the next several decades, particularly in Asian countries. Coal is relatively cheaper than other fossil fuels, making it an attractive option when other fuel prices increase. In addition, many countries possess large coal resources. Even with competition from petroleum and natural gas as energy sources, coal has acted as a moderate helper to meet the world energy demand Recent global warming and health hazards caused by micro-sized ultrafine particles prompted bans to the further utilization of coal.

Coal is a valuable asset that is abundantly stored within the Earth. Some suggestions propose the utilization of coal in its solid, liquid, and gaseous forms separately. Advocates of this approach argue that by doing so, coal resources can be optimally harnessed, ensuring maximum internal asset retention. However, economic considerations often lead to simpler methods of coal utilization, such as straightforward combustion for heat and electricity generation.

What will be the role of coal in the 21st century? Will its use be phased out or will it continue to play a basic role for certain countries and industries? And if so, for how long? With worsening climate change worldwide, CO_2 emissions related to coal usage will determine the fate of coal energy in the 21st century.

Although the International Energy Agency's (IEA) Net Zero by 2050 roadmap envisions the end of coal generation by 2040, there is a high probability that this timeline will not be met. Actual costs and commercial availability of alternative clean energy technology will surely have a detrimental effect on this scenario.

The primary areas of coal utilization in the present day include electricity generation, iron/steel blast furnaces, kilns for cement production, and certain high-temperature industries. Unlike historical uses for small-scale applications like house heating, modern coal utilization is concentrated in large-scale industrial processes. This circumstance might justify more in-depth technological efforts for improving efficiencies and disseminating best practices in the coal energy field.

While finding ways to eliminate or substantially reduce CO_2 emissions is crucial for the long-term sustainability of coal energy, the reality is that many developing countries, especially in Asia, heavily rely on coal for energy security. This dependence is likely to persist at least during the early part of the 21st century. As such, sharing information about ways to minimize CO_2 emissions and maximize process efficiencies while minimizing costs and environmental impacts should be encouraged.

Lessons from highly optimized experiences in developed countries, along with innovative statistical tools, are ready to be employed in the coal energy field. Leveraging these resources can enhance efficiency, reduce waste, and promote sustainable practices in coal utilization. This book discusses some of these issues.

The book contains six chapters in three sections on the past/future of coal energy, statistical tools application, and application technologies. The first section deals with the past and future of coal energy, ranging from fundamental questions on coal usage to the development history in coal mining. The second section includes two chapters that discuss the issue of improving productivity and effectiveness in the coal energy industry and the application of existing statistical tools. The third section presents examples of practical applications in coal utilization, such as gasification and high-value utilization of fly ash.

Chapter 1 critically deals with the pros and cons of coal energy with some philosophical discussion, while more focused on the detrimental impacts that have accumulated by coal mining as well as the future direction for environmentally benign coal use. Chapter 2 explains the history of and the equipment developed for enhancing productivity in coal mining as well as current clean coal technologies, including USC(Ultra-Supercritical)/IGCC(Integrated Gasification Combined Cycle) and recent CCUS(Carbon Capture Utilization and Storage) projects in Japan.

Chapter 3 presents an example of applying factor analysis in coal mining by extracting statistically significant variables from management/technology at seven Indian mines for resolving low productivity problems. Chapter 4 describes the Lean Six Sigma method for detecting which factors cause product failures and for seeking improvement suggestions in a rotary kiln case. Coal is still a major heating medium in rotary kilns used by the cement and smelting industries.

Chapter 5 explains the bubbling fluidized bed coal gasifier, providing information on basic fluidization fundamentals and examining practical pilot-scale details beneficial in university research labs. Chapter 6 demonstrates the utilization of waste coal fly ash, which can otherwise cause environmental problems for value-added materials like zeolites. It also summarizes existing literature data regarding synthesis methods and properties of zeolite products.

Some chapters in the book might not be considered novel scientific papers like those that typically appear in scientific journals. Instead, they are technical reports that share more in-depth, on-site information on the specific field of coal energy. This practical approach becomes significant when considering the widespread use of coal energy in existing facilities. These facilities often seek ways to maximize efficiency using tools that are readily available, along with external insights and on-site information that represent the best practices in the field.

I would like to thank all of the authors who shared their valuable experience and data on coal energy for this book. Thanks also to the staff at IntechOpen, particularly Publishing Process Managers Ms. Martina Scerbe and Ms. Ana Cink for all their help throughout the publication process.

Yongseung Yun
Institute for Advanced Engineering,
Yongin, Republic of Korea

Section 1

Past and Future of Coal Energy

Chapter 1

Perspective Chapter: The True Cost of Coal – Should Ego Veto Eco?

Oluwafikemi Iji

Abstract

There is an urgent need to address the ever-growing concerns about the long-term impacts of coal mining as a cheap energy source in the 21st century because the consequence of inaction threatens the health of the environment, which is inextricably linked to human health. The discovery of the world's largest solid fuel no doubt brought about industrial and modern technology revolution, but its cost on air, water, land, ecosystem, animal, and human health has brought about new realities that intreat urgent action. As humans, we sit on top of the food chain perpetuating our dominance over other species and studies have shown that with regards to the battle for the earth, human needs come first. To maintain a sustainable ecosystem, we need to foster a mutually beneficial relationship that promotes both the health and sustainability of our environment. A targeted transition away from coal to cleaner forms of energy will undeniably benefit the ecosystems, however appropriate measures are needed to continually reduce the environmental footprint, of the most available energy source so we can protect both the environment and human health.

Keywords: coal, mining, pollution, environment, global warming

1. Introduction

There are numerous economic and immediate benefits arising from coal mining [1] because it is a cheaper energy source and power drives the economy, hence, increased demand by many countries despite mounting evidence that using coal as an energy source poses serious threats to both humans, animals, and the environment, we share [2].

All phases of coal utilization involve many hazardous aspects [3]. Coal mining processes require major earth movements which disrupt and/or damage vegetation and topsoil, leaving mined areas prone to erosion. When coal is burned it also releases harmful pollutants into the atmosphere and this contributes to the climate change crisis, acid rain, and various respiratory health problems, and its left-over waste also poses the risk of water pollution and loss of habitat to wildlife [4, 5].

Coal is a non-renewable energy resource [6], which when mined, becomes difficult to clean up. It leaves an undesirable, long-term effect on the environment as many old mines are abandoned when it is no longer profitable to mine, desperately leaving the sites needing reclamation and/or remediation.

1.1 History of coal mining

The "black gold" as it was called when it was first discovered, is the most abundant fossil fuel on earth [7]. Coal mining, by definition, is the extraction of coal deposits from the surface of the earth and from underground. History showed that this energy source powered the Industrial Revolution. Large-scale exploitation of coal deposits was largely responsible for industrialization that occurred in that era: between the 18th–19th century.

Archeological evidence from the bronze age period from 3000 to 4000 years ago suggests the use of coal in funeral pyres in Wales. The Chinese earliest mentions of a coal mine being opened was over 3000 years ago and the Romans were the first recorded to have used coal more extensively; they used coal as fuel to heat baths, ornaments, iron forging, and for religious ceremonies and coal was used to worship the goddess of wisdom, Minerva [7, 8].

Around the 1700s due to the population boom, the demand for coal exploded. The invention of the steam engine helped the mining companies overcome the issue of flooded mines as water could be pumped out, which allowed coal mines to delve deeper into the ground [8]. The mines became not only deeper, but also more dangerous, with other threats arising from explosives, poisonous gases, and possible collapse of coal mines [9]. Nowadays, coal continues to be used directly (heating) and indirectly (producing electricity). It is used as fuel around the world, and in developing countries, it is the leading choice of energy [10].

Individual households and industrial furnaces use coal directly through burning. Widespread indirect use involves the use of coal to generate electricity [11]. Coal-fired power plants are arguably the most popular ways to produce and distribute electricity. Coal-fired power plants combust coal to power boilers whose steam is used to turn turbines to activate generators that produce electricity [12]. Likewise, the use of metallurgical coke for steelmaking has been a major driver for technological innovations [13]. Many countries generate electricity through coal-fired plants [14]. There is no denying the contribution and importance of coal to the world's energy budget. Its relative ease of extraction and ubiquitous nature, make it desirable for use as an energy source [15]. Coal production is not weather dependent, and upon mining it becomes very readily available for use all year round.

1.2 What is coal?

Coal is made up of trapped carbon and hydrogen-rich material, whose energy is released following combustion. Coal is a sedimentary rock that is usually either black or brownish-black in color [16]. Coal formation began hundredths of millions of years ago when plants, which were once alive, die with their stored solar energy acquired through photosynthesis [17]. The plants then began the process of decay. And this process of decomposing becomes intercepted for thousands of years, allowing for pressure and heat buildup on top of the plant remains and also for further physicochemical changes, which force oxygen out, such that its remains turn into coal [18].

Since coal is subjected to different phases of carbonization over millions of years, it is usually found at different stages of development [19]. There are several types of coal ranked according to how much it has changed over time. The general rule is that the deeper the coal seam, the more the temperature and pressure the coal becomes subjected to, and the greater its transformation into carbon [20].

1.2.1 Types of coal

Lignite - This coal is the lowest rank of coal. It contains low amounts of energy—its carbon content is about 25–35%. It comes from relatively young coal deposits. Lignite contains more moisture than other types of coal, usually crumbly brown and susceptible to fortuitous combustion.

Sub-Bituminous Coal - Contains more carbon than lignite, about 35–45%. In many parts of the world, sub-bituminous coal is considered "brown coal," along with lignite. Like lignite, sub-bituminous coal is mainly used as fuel for generating electricity.

Bituminous coal is formed under more heat and pressure and is usually between 100 and 300 million years old. It is named after the sticky, tar-like substance called bitumen and contains about 45–86% carbon. There are three types: smithing coal, coking coal, and cannel coal. Anthracite - This is referred to as the peak coal. With the most abundant carbon (97%), it contains the most energy. It is harder, denser, and glossier than other types of coal. It burns cleanly with very little soot and is used in water-filtration systems. (Source: Coal 101, Student Energy, www.studentenergy.org).

2. Coal form an ego standpoint

The worldwide consumption of coal as an energy source is quite extensive [14]. In South Africa 93% of electricity is generated by coal [21]. About 45% of the US national grid is coal driven while countries like China, India, Russia, and Australia are also heavily reliant on coal-powered energy [22]. The economics that drives the coal mining industry is huge, providing millions of people worldwide with jobs. Many of them have a wide range of knowledge and skills, ranging from miners to chemists, engineers, geologists, etc.

The coal industry is deemed critical to development in both the developed and developing world. Although as non-renewable energy, we cannot add to the world's coal, and it may not be available forever. However, with approximately 11 trillion tons of coal distributed worldwide, and recoverable coal estimated at 760 billion tons (Europe 49%, North America 29%, Asia 14%, Australia 6%, Africa, and South America 1% each), coal mining would still be around for a long time [7].

While coal by far is considered the dirtiest fossil fuel there is in terms of carbon emissions and air pollution [23], arguments to support its use as a way of life from a cultural and political standpoint continues to be made, despite evidence of the environmental cost and the health care cost [24], especially in comparison to other energy sources like natural gas and renewables like solar and wind.

In the US, some lawmakers continue to make the case for coal by appealing to people's sense of history and heritage. Blocking legislation that would speed the country's transition away from coal to other clean energy sources, and advocating to relax environmental legislations that hold them accountable for pollution, serves only the interests of the energy-producing companies [25]. The argument that the environmental cost of coal extraction pales in comparison to the economic toll, is one that many environmentalists find insupportable because they believe that this threat is beyond economic, and should, first and foremost, be a moral issue; an obligation owed to the planet.

Coal amongst other energy sources is a necessity for the developed world to maintain their prosperity and the developing ones to push to lessen poverty. It is pivotal to economic growth, infrastructural development and urbanization [26]. The demand for roads, airports and housing drives those of electricity, steel and cement typically rely on coal to fuel their production. Coal retains its prominence because of the global energy demand.

2.1 The energy-manipulating machine

There is no doubt that the world has largely become a consumerist society, especially in the developed countries where the means, the purchasing power, and the resources are replete.

Herbert Spencer (1820–1903) the British Philosopher, termed energy as being fundamentally responsible for material inequities between societies. The richer countries produce and have access to more energy than they need when compared to the poorer countries. There was the belief that a society that harnessed the most energy from its resources was by far superior [27].

Historically, Wilhelm Ostwald described energy transformation from its crude form into its useful form as the base for all social change "a guideline for cultural development" [28]. The race to exploit these natural resources was seen as a means of emancipation and empowerment [29–31]. so much so that even leaders of the communist revolution in Russia, thought electrification could be an agenda towards "true communism" [32]. The Germans on the other hand, through their engineers, were innovators in the field of electric power generation and transmission, they promoted the idea of surpassing profit with efficiency, making efficiency the primary regulator of the economy [33].

Whatever ideology drove massive energy production, profit was at its core, some ideologies even subscribed to the use of energy as a form of currency to replace gold [34]. Energy producers designed ways to market their products in a way that interacted with societies, cultures, economics, and the environment [35]. The societies measured value based on what was owned; soon the race to possess numerous gadgets which required power to run, became a fad. Unaware, the public was successfully programmed into making consumption a way of life.

In Germany for example, there was an economic boom post-WWII, and increasing the sales of electricity became part of the social construct for nation-building, due to mass consumption. An assortment of electricity advertisements marketed modernity, freedom, leisure, and progress as their subtexts. France on the other hand, marketed a new social order reinforced by electricite de France, promoting a rapid accumulation of appliances in households, for comfort and then later, as lifestyle objects [36].

Many high-energy consuming countries were not formed through cultural exemplifications of energy alone, but also, by the actual consumption patterns. Electricity utility companies envisioned households as a part of their production and marketing strategy and used appliances to shape consumer practices.

2.2 Making a case for coal

There's undoubtedly, a huge skepticism when it comes to clean coal and quite frankly, it sounds paradoxical to use the label 'clean coal,' considering how inherently dirty and destructive the mining process is, from extraction to utilization. There has been a concerted effort since the first coal-fired generator began running in England back in the late 1800s to find cleaner ways to burn coal and this continues to develop.

Progress has been made from burning coal inside homes to moving power plants with tall smokestacks outside of cities, thus, reducing the immediate problem of inhaling soot. Further advancement came with the technology of installing filters to retain particulates and scrubbers to trap the sulfur gases that were generating acid rain [37]. The newest technological advances are now able to capture CO2, preventing its release into the atmosphere. The trace metals and sulfur are concentrated as coal waste stored in slurry ponds or landfills as fly ash and gypsum. However, occasionally due to flooding or procedural mishap, some of this waste spill into rivers.

Although the older coal power plants were more notorious for releasing pollutants, some people continue to make the case that deploying higher-tech plants with lower emissions, such as the ultra-supercritical plants, which also have lower operational and maintenance costs technologies may be beneficial, the downside is that they are more capital intensive [38]. In general, as operations become more advanced in tech, they become cleaner and more efficient, and require less human involvement, thus reducing potential hazards to miners and the workforce.

Advocating for stricter emission standards has been the view of many environmental stakeholders. In China, for example, there has been a steady tightening of pollution emissions standards as Beijing rolled out a new air quality monitoring system that provides real-time information on air quality across the nation [39], pressured by its citizens, authorities abandoned the previously subjective air quality reporting system [40]. Currently, China's new air pollution emissions standards for new and existing coal-fired power plants are sturdier than those from the European Union and the United States [41].

One of the strongest advocacies for the continued interest in coal mining has been that the world's economy may collapse if we move away since power drives the economy. The livelihoods of many people depend on the extractive industries and even for some, it is an identity and a way of life. China's newer power plants are becoming more efficient in terms of labor, employing fewer human hands [42]. The coal sector in China was expected to lay off approximately 1.3 million workers between 2016 to 2020 [43]. Similar pattern is recounted in the US where coal mining employment has been experiencing a steady decline [44–46]. A shift from labor-intensive underground mining in the East to highly mechanized surface mining in the West is responsible for the decline in employment [47]. Other associated factors are the falling demand for coal and electricity, price drop for natural gas, solar, and wind which serve as alternative renewable sources, and the rising energy efficiency [48].

The simple fact shows that in the United States new jobs are being added in the renewable energy sector [49], which competes favorably with coal-fired electricity generation—particularly shale gas, wind, and solar. Similarly, U.S. electric utility executives are directing their business models towards renewable options—and away from coal [50]. Some visionary leaders are making strategic choices in the direction they want their country to go especially in China, where the political will is rooted in both international and domestic pressure, however just like in the rest of the world, vested interests serve as a political clog in the wheels of accelerated progress.

Coal remains abundant and cheap as an energy source, the question then arises if the fact justify the true cost paid by the people whose lives are profoundly altered through exposure to hazardous materials, or the communities who breathe air polluted by coal or mining dust, or drink water contaminated by coal mines or coal combustion residues [51–53]. The scale of pollution that coal generates at every step when compared to other sources make it an environmental liability. If new technologies were adequate to remove air and other forms of pollutants from coal, if CO2 were captured and stored safely in underground reserves, perhaps there would be less call to shift from fossil fuels to alternative sources.

3. Coal form an eco-standpoint

One of the most powerful tools for many fossil fuel energy companies has been the power of denial. The intentional and persistent denial of the link between their activities and the climate change crisis. There is an alleged campaign of deception and misinformation which has been since the early 1990s. The World Coal Institute's position was that: "model-based projections are controversial, uncertain, and without confirmation, and that scientists were divided in their opinion about the likelihood and consequences of climate change" [54].

With almost 40% of the world's carbon dioxide emissions arising from burning coal, the environmental effects of the coal industry seem harmful, producing gases such as carbon monoxide (CO), sulfur dioxide (SO2), sulfur trioxide (SO3), nitric oxide (NO), and nitrogen dioxide (NO2) [55, 56]. A report suggested that considering the severe damaging effects on human health and environmental health, that it needs to be addressed as a public health issue [57]. Occupational hazards to miners resulting from accidents and also to their health are of serious concern, an example is black lung disease (coal workers' pneumoconiosis) [58].

Although many corporations are largely not held accountable for the damage they create, by releasing lead, mercury, selenium, and other toxicants into the environment, this raises serious moral and ethical issues because it puts humans and the planet at risk. At the least, these threats should trigger precautionary measures even if direct cause and effect correlation has not been fully elucidated [59]. The onus probandi then lies with the coal industry that no harm is caused [60].

3.1 Environmental concerns and consequences

3.1.1 Land

The environmental impacts of coal mining are dramatic. Due to major earth movements, the landscape is torn apart, destroying habitats and entire ecosystems. Mining also causes landslides and subsidence (when the ground begins to sink or cave in), thus reducing the value of the natural environment in the surrounding land [61], or rendering the land completely unusable.

Vegetation, wildlife, and habitat, general topography are usually sacrificed in areas subjected to mining [62]. Paleontological, cultural, and archeological loss may ensue due to the blasting, ripping, and excavation processes [63]. Heavy hauling of soil and coal increases dust dispersal during mining operations which degrade air quality and impacts vegetation. Waste piles accumulate, and there is a loss of topsoil, forming large infertile wastelands [64]. The whole sight, sound, and smell of a mined area may become affected.

3.1.2 Water

Coal mining activities can constitute a major and persistent source of pollution to water bodies either directly or indirectly. This presents harmful consequences to aquatic animals, poses risks to humans who are downstream end users, hinders aquatic activities, and impairs the quality of water for use [65]. Other changes may affect the physicochemical and biological characteristics of the water bodies in a manner that potentially threatens the organisms inhabiting the water, the aquatic ecosystems, and biodiversity [66] which may then upset ecological and ecosystem services, disturbing its quality and productivity.

One major way coal mining activities usually contribute to water pollution is by promoting an acid mine drainage environment. It happens when chemical changes occur from weathering of minerals (pyrite) which contacts water and air, then discharges with the resultant release of metals, generating acidity (sulfuric acid) to receiving waterbodies [67]. Waste piles also contribute to the deterioration of water quality, groundwater contamination, toxic trace elements, and the presence of highly dissolved solids. Coal-fired boilers, using either coal or lignite rich in limestone also contribute to salinity in water bodies [68].

3.1.3 Air

Burning coal releases gases and particulates that are harmful to the environment. Hazardous air pollutants emitted from coal-fired plants include toxic heavy metals (e.g., As, Pb, Cd, Se, Hg), sulfur dioxide, nitrogen oxides, carbon monoxide, and radioactive elements (e.g., uranium, radium, thorium) [69].

Particulates (fly ash), and fugitive dust released are a risk to humans, animals, and the environment. Coal dust are respiratory irritants that have been linked to black lung disease, chronic obstructive pulmonary disease, and silicosis [70]. Likewise, the potential health impacts of nanoparticles generated during coal mining and coal combustion continue to raise questions [71], and reports of increasing human morbidity and mortality in people living near and far from coal power plants owing to air pollution are disturbing [72].

The extraction and burning of coal releases sequestered carbon into the atmosphere, which leads to a build-up of greenhouse gases responsible for climate change and global warming [73]. Coal fires emit tons of greenhouse gases, and due to their combustible nature, coal seams can self-ignite and burn underground threatening valuable infrastructures [74]. Lightning, wildfires, and explosives used in the extraction process may also start coal fires which release deleterious gases.

3.2 Global warming and climate change

Greenhouse gases produced by human activities have been implicated in global warming and rising climate-related surface level air warming [75] which occurs at the rate of 0.8 to 1.3°C, and this assertion is unequivocal. Burning fossil fuel releases large amounts of carbon into the atmosphere, causing CO2 levels in the atmosphere to rise along with CH4 and to a lesser extent, NO [76]. The observed impact within the scientific community is shifting from that of concern to threats.

Changing weather patterns are significantly changing the timing of the seasons and affecting species and their ecology [77]. The disappearance of glaciers is forcing migration towards the poles to maintain temperature stability. A study showed that since the start of the twentieth century, animals' reproduction and migration occurred earlier than they did when the average global temperature was 1°F cooler [78]. Scientists observed the earlier arrival of migrant birds, the earlier appearance of butterflies, earlier spawning in amphibians, and the earlier flowering of plants [79].

Evidence suggests a broad range of organisms with different geographical distributions have been affected by climatic changes.

Climate change issues have resulted in food and water shortages, and increased the risk of flooding, particularly for those living in poverty [80], and countries that cannot afford the infrastructures to accommodate floods, droughts, severe storms, and epidemics of disease that make them vulnerable. It is projected that by 2065, the community of property reinsurers' losses from climate impacts on the global economy may tend towards bankruptcy if we do not act now to mitigate the risks [81]. Extremes of temperature are reported to account for 9.4% of global deaths (up to 5 million annually) between 2000 and 2019 [82].

4. Human health impact

4.1 Occupational hazard

Historically, coal mining is thought to be a very dangerous activity, with attending historical disasters arising from wall failures and vehicle collisions, the danger of suffocation, gas poisoning, and explosions. Many miners die annually, either directly from accidents in coal mines or adverse health consequences. The United States reported an average of 23 coal miners' death per year from 2007 to 2016, while Chinese miners related death around the same time was 6027 [83, 84].

Other risks are those associated with suffocation due to the buildup of hazardous gases such as carbon dioxide and nitrogen, the risk of explosion from highly flammable gases, and asphyxiation from possible methane poisoning [85].

4.2 Pathophysiology of coal inhalation in humans

4.2.1 Lungs

The lungs are the predilection site for air pollutants [86].Coal inhalation triggers several respiratory diseases collectively known as coal mine-dust lung disease which affect the lung health of miners; Silicosis, chronic bronchitis, chronic obstructive airway disease, including emphysema and mixed dust pneumoconiosis [87].

Indirect exposure predisposes to chronic obstructive pulmonary disease (COPD), asthma, lung cancer, and respiratory infection, children and the elderly are more vulnerable [88], leading to reduced life expectancy and co-morbidity.

Black lung disease is still common in some mining countries, affecting 4% of US workers and 0.2% percent of China workers. Rates may be higher than reported in some regions, continued prevalence is projected in the future [89].

4.2.2 Heart

There is a strong correlation between increased risks of coronary artery disease, heart failure, stroke, cardiovascular morbidity and mortality, and environmental air pollutants [90]. Many other cardiovascular diseases have been specifically linked to contaminants arising from coal and coal processing found in air and water [91].

The pathogenesis of these cardiac diseases is hypothesized to be related to the development of oxidative stress, immune-mediated systemic inflammation, and endothelium dysfunction, resulting in hypertension, cardiac hypertrophy, atherosclerosis, thrombosis, and myocardial ischemic damage [92] which affect the quality of life and could lead to mortality.

4.2.3 Brain

The human brain is also a likely target for pollutants released from coal burning. It causes brain structural changes, and neuropsychological change alters cognitive development and promotes memory impairment, especially in children [93]. Neurotoxic pollutants are present in coal; As, Hg and Al trigger oxidative stress and apoptosis in the central nervous system [94].

People exposed to arsenic present with axonal degeneration of peripheral nerve and the role of Hg in neurodegenerative diseases, like Alzheimer's disease has been elucidated, and Hg is reported to be the most substantial trace-element causing an imbalance in the amygdala, hippocampus, and cerebral cortex [95].

Other associated disorders related to coal inhalation include causing possible DNA damage [96], drop in reproductive capacities and infertility [97], fetal- toxicity, and other co-morbidiites [98].

5. The future coal field

Although coal mining may have earned itself a bad name especially amongst environmentalist, and scientists who have clearly established a link between climate change and coal burning [99] and its potentially dangerous consequences for humanity [100]. Our current reality does not support a total abandonment from coal use, and despite these issues, it is crucial to explore ways to reduce anthropogenic emissions by re-inventing coal utilization for the future. For example, the use of technology for Carbon Capture and Storage (CCS) or Carbon Capture and Sequestration (CCS), will help reduce greenhouse gas emissions [101], likewise, technological improvements that employ supercritical (SC) and ultra-supercritical (USC) technologies make coal-fired power plants become more efficient and cleaner [102], because it produces smaller amounts of CO2 per unit of electricity generated, and consequently, require less to be captured, this ultimately improves efficiency and cost, and lowers penalties for CO2 release.

Moreover, the fact that in some parts of the world, coal is still the primary source of energy for electricity generation and transitioning to other sources may be costly and difficult, efforts must therefore be geared towards improving fuel combustion efficiency and pollution control equipment performance. Similarly providing much needed funding for equipment upgrading, performance monitoring and diagnostic testing would be immensely beneficial to curbing pollution [103].

Offering incentives to continue research and development towards a near-zero emission technology can further help make the case for the continued use of the world's most abundant solid fuel [104]. The integrated gasification combined cycle (IGCC) is said to have the potential to achieve higher efficiency, it is intended to decarbonize coal combustion and these system technologies when refined, can be deployed for commercialization purposes [105].

6. Conclusion

Coal as an energy source has contributed tremendously to industrialization and technological advancement and the argument is not to shut down all coal plants by tomorrow, but to acknowledge its setbacks as a power-generating source that competes with the essential elements necessary for human survival: water and air. That in mind should be a driver towards being intentional about putting adequate measures in place to preserve human health and protect the environment.

Unfortunately, neither the producers nor consumers of coal energy are immune from the consequences of the danger the industry poses. The realization that we are all a part of this increasingly vulnerable community, should prompt a new sense of common purpose, that sets the tone for a renewed relationship. The energy conglomerates want to thrive and promote their business, we the consumers understand energy is a necessity for a vibrant economy, however, their activities and our demand should not be solely motivated by the economics of the market which does not account for community and/or environmental consequences of its operations.

There must be a way to expand wealth in a global context without destroying the physical environment on which it depends. Humans have fared well when faced with challenging situations. We have negotiated peace in the time of war, survived and beat a global pandemic, and there is hope that we can find a way to avert this impending and inevitable catastrophe. The entire human enterprise collectively needs to see transitioning to sustainable energy as a global project that could pull us out of the impending natures fury.

Nowadays, alternative fuels (e.g., natural gas, wind energy and solar power) are becoming widely available for generating power [106] and more investments into alternative sources, offer a greater chance of rendering coal mining in some countries uneconomic, especially the older mines. However, the political will to support such projects will be needed. The recent breakthrough announced by US scientists in nuclear fusion [107] involves the technology that could potentially provide near-limitless clean energy, but this development is yet to come to the fore in any major way.

There is optimism for the future in our resourcefulness and creativity to tackle this problem, but time may not be on our side as people around the globe continue to feel the economic impact of climate change and the cost of mitigation becoming unjustifiable.

In conclusion, although technological advancements in the future could make coal utilization cleaner and more efficient, we need to do more in the present. It is worthy of note that cost should not be the reason why the world has not committed to clean energy despite the evidence that millions of jobs will be created, and that the industry is growing faster than fossil fuel-powered industries. Our biggest impediment is change — one that "big coal" does its best to encourage.

Author details

Oluwafikemi Iji
Federal College of Animal and Production Technology, Institute of Agricultural Research and Training, Moor Plantation, Apata, Ibadan, Oyo State, Nigeria

*Address all correspondence to: fikemi@yahoo.com

References

[1] Baiyewu-Teru A. The History of Coal in Nigeria. Abuja, Nigeria: Heinrich-böll-stiftung; 2015. Available from: https://ng.boell.org/en/2015/12/03/history-history-coal-nigeria. Accessed 1 December 2022

[2] Union of Concerned Scientists (2008, updated 2017). Coal and Air Pollution. Available from: http://www.ucsusa.org/clean_energy/coalvswind/c02c.html [Accessed: December 7, 2022]

[3] Chadwick MJ, Highton NH, Lindman N. Occupational and Environmental Health Problems from Coal Utilization and Conversion, Environmental Impacts of Coal Mining & Utilization. Oxford: Pergamon press; 1987. pp. 245-281

[4] US Energy Information Administration – (EIA). Independent Statistics and Analysis. Coal Explained: Coal and the Environment. https://www.eia.gov/energyexplained/coal/coal-and-the-environment.php

[5] Mitchell C. Environmental Impacts of Coal Mining. 2022. Available from: https://worldanimalfoundation.org/advocate/wildearth/params/post/1285440/impacts-of-coal-mining [Accessed: December 7, 2022]

[6] Turgeon A & Morse E. Coal. 2022. Available from: https://education.nationalgeographic.org/resource/coal [Accessed: December 5, 2022]

[7] Ramani RV & Evans MA. Coal Mining. Encyclopedia Britannica. 2020. Available from: https://www.britannica.com/technology/coal-mining [Accessed: December 7, 2022]

[8] Ghosh TK, Prelas MA. Coal. In: Science S, editor. Energy Resources and Systems: Fundamentals and Non-Renewable Resources. Vol. 1. Columbia, USA: Springer Science Business Media; 2009. pp. 159-279

[9] Olney M. A Brief History of Energy – Coal. 2022. Available from: https://www.dyballassociates.co.uk/a-brief-history-of-energy-coal [Accessed: November 23, 2022]

[10] International Energy Agency. World Energy Outlook. Paris: IEA; 2016. Available from: https://www.iea.org/reports/world-energy-outlook-2016, License: CC BY 4.0

[11] World Coal Association Coal & electricity. 2016. Available at: https://www.worldcoal.org/coal/uses-coal/coal-electricity [Accessed: April 20, 2016]

[12] Britannica. The Information Architects of Encyclopaedia. "coal mining". Encyclopedia Britannica. Available from: https://www.britannica.com/facts/coal-mining [Accessed: November 23, 2022]

[13] Kennedy C. The energy embodied in the first and second industrial revolutions. Journal of Industrial Ecology. 2020;**24**(4):887-898

[14] Zou C, Zhao Q, Zhang G, Xiong B. Energy revolution: From a fossil energy era to a new energy era. Natural Gas Industry B. 2016;**1**:1-11

[15] Houser T, Bordoff J & Masters P. The State of Coal in the United States. Columbia University Center on Global Energy Policy. 2018. Available from: https://energypolicy.columbia.edu/research/report/state-coal-united-states

[16] Rađenović A. Inorganic constituents in coal. Kemija u Industriji. 2006;**55**(2):65-72

[17] Hower JC, Rimmer SM, Mastalerz M, Wagner NJ. Notes on the mechanisms of coal metamorphism in the Pennsylvania anthracite fields. International Journal of Coal Geology. 2019;**202**:161-170

[18] Dai S, Bechtel A, Eble CF, Flores RM, French D, Graham IT, et al. Recognition of peat depositional environments in coal: A review. International Journal of Coal Geology. 2020;**219**:103383

[19] O'Keefe JMK, Bechtel A, Christanis K, Dai S, Di Michele WA, Eble CF, et al. On the fundamental difference between coal rank and coal type. International Journal of Coal Geology. 2013;**118**:58-87

[20] Tarafdar A, Sinha A. Polycyclic Aromatic Hydrocarbons (PAHs) Pollution Generated from Coal-Fired Thermal Power Plants: Formation Mechanism, Characterization, and Profiling. In: Agarwal R, Agarwal A, Gupta T, Sharma N, editors. Pollutants from Energy Sources. Energy, Environment, and Sustainability. Singapore: Springer; 2019

[21] Oberschelp C, Pfister S, Raptis CE, Hellweg S. Global emission hotspots of coal power generation. Nature Sustainability. 2019;**2**:113-121

[22] Rauner S, Bauer N, Dirnaichner A, Dingenen RV, Mutel C, Luderer G. Coal-exit health and environmental damage reductions outweigh economic impacts. Nature Climate Change. 2020;**10**:308-312

[23] Liu K, Wang S, Wu Q, Wang L, Ma Q, Zhang L, et al. A highly resolved mercury emission inventory of Chinese coal-fired power plants. Environmental Science & Technology. 2018;**52**(4):2400-2408

[24] Oliveira MLS, da Boit K, Pacheco F, Teixeira EC, Schneider IL, Crissien TJ, et al. Study of coal cleaning rejects by FIB and sample preparation for HR-TEM: Mineral surface chemistry and nanoparticle-aggregation control for health studies. Journal of Cleaner Production. 2018;**188**:662-669

[25] Bogus law D. Joe Manchin's dirty empire. The Intercept. 2021. Available from: https://theintercept.com/2021/09/03/joe-manchin-coal-fossil-fuels-pollution/ [Accessed: December 10, 2022]

[26] International Energy Outlook 2007. (IEO) of May 2007. Energy Information Administration. Office of Integrated Analysis and Forecasting U.S. Department of Energy Washington, DC. Available from: www.eia.doe.gov/oiaf/ieo/index.html

[27] Spencer H. Geddes Shared his Admiration of Electricity as an Agent of Societal Betterment with Many Others. First Principles. New York: Burt AL. 1880. p. 1880

[28] Ostwald W. Energetic Foundations of Cultural Studies. Leipzig: Klinkhardt; 1909

[29] Pursell CW Jr. Domesticating modernity: The electrical Association for Women, 1924-1986. British Journal for the History of Science. 1999;**32**(1):47-67

[30] Cottrell, Frederick W. Energy and Society: The Relation between Energy, Social Changes, and Economic Development. New York: McGraw-Hill Book Company, Inc.; 1955. pp. 330

[31] Hughes TP. American Genesis: A Century of Invention and Technological Enthusiasm. New York: Viking; 1989. pp. 353-381 Nye: Electrifying America, 304-35

[32] Coopersmith J. The Electrification of Russia, 1880-1926. On Lenin and Soviet

Electrification. Ithaca, NY: Cornell University Press; 1992. p. 1992

[33] Hellige HD. Economic, energy and material cycles in a secular way perspective: From thermodynamic disenchantment of the world to recycling-oriented growth society. In: Hübinger G, Schulin E, editors. Universal History and National Histories. Rombach: Freiburg im Breisgau; 1994. pp. 291-315

[34] Zachmann K. Efficiency versus value: On the emergence of the conflict between the technical and the economy "concept of economics". History of Technology. 1995;**62**(2):103-131

[35] Howard Scott's Own, Infamous "Hotel Pierre" Address of January "Technocracy". 1993. Available from: http://www.technocracy.org/technical-alliancetn/howard-scott/208-societyscott [Accessed: December 8, 2022]

[36] White LA. Energy and the evolution of culture. American Anthropologist. 1943;**43**(3):335

[37] Bernhardt ES. What does 'Clean Coal' Mean, and Can It Save the Planet? 2017. Available from: https://www.pbs.org/newshour/science/column-clean-coal-mean-can-save-planet [Accessed: December 12, 2022]

[38] U.S. Energy Information Administration. Capital Cost Estimates for Utility Scale Electricity Generating Plants (U.S. Department of Energy). 2016. https://www.eia.gov/analysis/studies/powerplants/capitalcost/pdf/capcost_assumption.pdf

[39] Jinran Z. Monitoring Network to be Further Expanded. China Daily. 2017. http://www.chinadaily.com.cn/china/2017-04/07/content_28827498.htm

[40] Xinhua. Beijing No Longer Counting 'Blue Sky days. ChinaDaily. 2012. Available from: http://usa.chinadaily.com.cn/china/201206/06/content_15476803.htm

[41] Guo Jia Fa Zhan Gai Ge Wei (National Development and Reform Commission), Huan Jing Bao Hu Bu (Ministry of Environmental Protection), and Guo Jia Neng Yuan Ju (National Energy Administration). Mei dian jie neng jian pai sheng ji yu gai zao xing dong ji hua (2014-2020 nian) (Coal-Fired Energy-Savings, Emissions Reduction, Upgrade, and Transformation Action Plan (2014-2020)). 2014. Available from: http://www.mep.gov.cn/gkml/hbb/gwy/201409/W020140925407622627853.pdf.

[42] Waigaoqiao No. 3 Power Station Leadership Team, Interviews with Authors, Shanghai, December 2016. For more about Waigaoqiao and its Technology, see Thomas W. Overton. Top Plants: Shanghai Waigaoqiao No. 3, Shanghai, China. POWER. 2015. Available from: http://www.powermag.com/shanghai-waigaoqiao-no-3-shanghai-china/?pagenum=1

[43] Winter Nie. Here's Why China Laying Off 1.8 Million Workers Is Actually Good News. Fortune. 2016. Available from: http://fortune.com/2016/03/03/china-mass-layoffs-steel-coal/

[44] Mine Safety and Health Administration, Coal Employment by State and County (CY 2009-2015) (U.S. Department of Labor). Available from: https://www.msha.gov/sites/default/files/Data_Reports/Charts/Coal_Employment_by_State_and_County_CY09to15.pdf [Accessed: April 2017]

[45] Personal Communication from Mine Safety and Health Administration Regarding its Forthcoming Coal Employment and Production Report, CY 2012 to 2016. U.S. Department of Labor. Available from: www.msha.gov. 2017

[46] U.S. Energy Information Administration. Today in Energy: Quarterly Coal Production Lowest Since the Early 1980s. 2016. Available from: https://www.eia.gov/todayinenergy/detail.php?id=26612.

[47] U.S Department of Energy. Quadrennial Energy Review: Transforming the Nation's Electricity System: The Second Installment of theQER. 2017.Available from: https://energy.gov/sites/prod/files/2017/02/f34/Quadrennial%20Energy%20ReviewSecond%20Installment%20%28Full%20Report%29.pdf

[48] U.S. Energy Information Administration. Today in Energy: Natural Gas Prices in 2016 were the Lowest in Nearly 20 Years. 2017. Available at https://www.eia.gov/todayinenergy/detail.php?id=29552.

[49] U.S. Department of Energy, U.S. Energy and Employment Report. 2017. Available from: https://energy.gov/sites/prod/files/2017/01/f34/2017%20US%20Energy%20and%20Jobs%20Report_0.pdf

[50] Fu S, Renick O. AEP's CEO Weighs in on Trump's Energy Plan. New York, USA: Bloomberg; 2017. Available from: https://www.bloomberg.com/news/videos/2017-03-28/aep-s-ceo-weighs-in-on-trump-s-energy-plan-video

[51] US National Library of Medicine. Medline Plus.Available from: https://medlineplus.gov/ency/article/000130.htm [Accessed: December 29, 2022]

[52] Bernhardt ES, Lutz BD, King RS, Fay JP, Carter CE, Helton AM, et al. How many mountains can we mine? Assessing the regional degradation of central Appalachian rivers by surface coal mining. Environmental Science & Technology. 2012;**46**:8115

[53] Hendryx M, Ahern MM. Relations between health indicators and residential proximity to coal mining in West Virginia. American Journal of Public Health. 2008;**98**(4):669

[54] E-coal. World Coal Institute Briefing no. 7. New York: INC 5; 1992

[55] Orem H, Finkelman RB. Coal formation and geochemistry. Treatise on Geochemistry. 2003;**7**:191-222

[56] Munawer ME. Human health and environmental impacts of coal combustion and post-combustion wastes. Journal of Sustainability Mining. 2018;**17**:87-96

[57] Union of Concerned Scientists. Environmental Impacts of Coal Power: Air Pollution. Reports & multimedia/explainer. Retrieved 9 December 2022. Abelard.org (2018 January 13) Way back Machine, Fossil fuel disasters. 2005

[58] Wingspread Statement on the Precautionary Principle. Available from: http://www.gdrc.org/u-gov/precaution-3.html [Accessed: December 12, 2022]

[59] Gilbert SG. Public Health and the Precautionary Principle, Northwest Public Health. Washington, D.C USA: University of Washington School of Public Health & Community Medicine, Spring/Summer; 2005. Available from: https://www.tpchd.org/files/library/427d8e6368fb4b6c.pdf [Accessed: 12 December 2022]

[60] Hamilton MS. Mining environmental policy: Comparing Indonesia and the USA. In: Ashgate Studies in Environmental Policy and Practice. Burlington, VT: Ashgate Publishing; 2005 ISBN 978-0-7546-4493-4

[61] U.S. Department of the Interior. Permanent Regulatory Program

Implementing Section 01(b) of the Surface Mining Control and Reclamation Act of 1977: Environmental Impact Statement. Washington, D.C; U.S. Department of the Interior; 1979

[62] United States Department of the Interior Coal. Construction and mining impacts. In: The Wayback Machine. Washington, D.C.: Office of Indian Energy and Economic Development, Tribal Energy and Environmental Information Clearinghouse (TEEIC); 2012. Retrieved 9 December 2022

[63] Nordquist M, Nandan SN. United Nations Convention on the Law of the Sea 1982, Volume VII: A Commentary. Vol. 7. Leiden, US: Martinus Nijhoff Publishers; 2011

[64] Wilhelmsson D, Thompson RC, Holmström K, Lindén O, Eriksson-Hägg H. Managing ocean environments in a changing climate: Sustainability and economic perspectives. Marine Pollution. 2013;**74**:127-169

[65] James TI. Water Quality of Streams Draining various Coal Measures in the North-Central West Coast. 2003. Available from: www.wcrc.govt.nz/council/publications/Science%20 Reports/AMD%20report%20 AusIMM%20Conference%20Paper.pdf. [Accessed January 21, 2023]

[66] Keating M. Cradle to the Grave: The Environmental Impacts from Coal. 2001. Available from: www.catf.us/publications/reports/Cradle_to_Grave.pdf [Accessed: 9 December 2022]

[67] Union of Concerned Scientists (2008). Coal and Air Pollution. Union of Concerned Scientists, Reports & Multimedia, Cambridge, MA. Article dated 28 July 2008 (updated 19 December 2017). Available from: https://www.ucsusa.org/resources/coal-and_air_pollution. Accessed 11 December 2020

[68] Miller BG. The effect of coal usage on human health and the environment (chapter 4). In: Miller BG, editor. Clean Coal Engineering Technology. Boston, MA: Butterworth-Heinemann; 2011. pp. 85-132

[69] NIOSH [National Institute for Occupational Safety and Health]. Coal Mine Dust Exposures and Associated Health Outcomes: A Review of Information Published since 1995. Current Intelligence Bulletin. Vol. 64. Washington, DC: Department of Health and Human Services; 2011. p. 56. Available from: https://www.cdc.gov/niosh/docs/2011172/pdfs/2011-172.pdf

[70] Silva LFO, Pinto D, Dotto GL, Hower JC. Nanomineralogy of evaporative precipitation of efflorescent compounds from coal mine drainage. Journal of Geoscience Frontiers. 2021;**12**(6):101003

[71] Saikia BK, Saikia J, Rabh S, Silva LFO, Finkelman R. Ambient nanoparticles/nano minerals and hazardous elements from coal combustion activity: Implications on energy challenges and health hazards. Geoscience Frontiers. 2017;**9**:863-875

[72] IEA [International Energy Agency]. Global Energy & CO_2 Status Report. Paris: IEA; 2019. Available from: https://www.iea.org/reports/global-energy-co2-status-report-2019

[73] International Energy Agency, Paris, France. p. 29. Available from: https://www.iea.org/reports/global-energy-co_2-statusreport-2019 [Accessed: December 12, 2020]

[74] David D. Unseen Danger: A Tragedy of People, Government and the Centralia Mine Fire. Pennsylvania, USA: University

of Pennsylvania Press; 1986 ISBN 978-0-8122-8022-7

[75] IPCC (Intergovernmental Panel on Climate Change). In: Masson-Delmotte V, Zhai P, Pirani A, Connors SL, et al., editors. Climate Change 2021: The Physical Science Basis (PDF). Contribution of Working Group I to the Sixth Assessment Report of the Intergovernmental Panel on Climate Change. Cambridge, England: Cambridge University Press; 2021

[76] US Environmental Protection Agency (EPA) Quote from public-domain source. Causes of Climate Change: The Greenhouse Effect Causes the Atmosphere to Retain Heat. Washington, D.C., USA: EPA; 2012. Archived from the original on 8 March 2017. Retrieved 11 December 2022

[77] Rice D. Coal Is the Main Offender for Global Warming, and Yet the World Is Using it more than Ever USA Today, Nation (Section). Tysons Corner, VA: Gannett Company; 2019 Article dated 26 March 2019. Available from: https://www.usatoday.com/story/news/nation/2019/03/26/climate-change-coal-still-king-global-carbonemissions-soar/3276401002/

[78] Root TL et al. Fingerprints of global warming on wild animals and plants. Nature. 2003;**421**:57-60

[79] Parmesan C, Yohe GA. Globally coherent fingerprint of climate change impacts across natural systems. Nature. 2003;**421**:37-42

[80] Billions face climate change risk". BBC News Science/Nature. 2007. Availble from: http://news.bbc.co.uk/2/hi/science/nature/6532323.stm. Retrieved 12 December 2022

[81] The Swiss Re Group: World Economy Set to Lose up to 18% GDP from Climate Change If no Action Taken, Reveals Swiss Re Institute's Stress-Test Analysis. Available from: https://www.swissre.com/media/press-release/nr-20210422-economics-of-climate-change-risks.html. Achieved 22 Apr 2021. Retrieved 11 December 2022

[82] Zhao Q et al. Global, regional, and national burden of mortality associated with non-optimal ambient temperatures from 2000 to 2019: A three-stage modelling study. The Lancet Planetary Health. 2021 July 1;**5**(7):415-425

[83] Mine Safety and Health Administration. Coal Mining Fatalities by State by Calendar Year 2017. U.S. Department of Labor. Available from: https://www.cdc.gov/niosh/mining/data/default.html. Archived from the original (PDF) on 23February 2011. Retrieved 12 December 2022

[84] Deconstructing Deadly Details from China's Coal Mine Safety Statistics. Available from: https://web.archive.org/web/20070930122719/http://www.clb.org.hk/public/contents/news?revision_id=19324&item_id=19316.

[85] Gupta AP, Dutta AK, Basu R. Coal basically a hazardous industry and source for numerous occupational diseases for workers. IJRDO. 2019;**5**(1):1-18

[86] Marino E, Caruso M, Campagna D, Polosa R. Impact of air quality on lung health: Myth or reality? Therapeutic Advances in Chronic Disease. 2015;**6**(5):286-298

[87] Perret JL, Plush B, Lachapelle P, Hinks TS, Walter C, Clarke P, et al. Coal mine dust lung disease in the modern era. Respirology. 2017;**22**(4):662-670

[88] Kurt OK, Zhang J, Pinkerton KE. Pulmonary health effects of air pollution. Current Opinion in Pulmonary Medicine. 2016;**22**(2):138-143

[89] Blackley DJ, Halldin CN, Laney AS. Continued increase in prevalence of coal workers' pneumoconiosis in the United States, 1970-2017. American Journal of Public Health. 2018;**108**(9):1220-1222

[90] Hendryx M, Higginbotham N, Ewald B, Connor LH. Air quality in association with rural coal mining and combustion in new South Wales Australia. The Journal of Rural Health. 2019;**35**(4):518-527

[91] Hendryx M, Zullig KJ. Higher coronary heart disease and heart attack morbidity in Appalachian coal mining regions. Preventive Medicine. 2009;**49**(5):355-359

[92] Aung N, Sanghvi MM, Zemrak F, Lee AM, Cooper JA, Paiva JM, et al. Association between ambient air pollution and cardiac morpho functional phenotypes: Insights from the UK biobank population imaging study. Circulation. 2018;**138**(20):2175-2186

[93] Alvarez-Pedrerol M, Rivas I, Lopez-Vicente M, Suades-Gonzalez E, Donaire-Gonzalez D, Cirach M, et al. Impact of commuting exposure to traffic-related air pollution on cognitive development in children walking to school. Environmental Pollution. 2017;**231**(1):837-844

[94] Mochizuki H. Arsenic neurotoxicity in humans. International Journal of Molecular Sciences. 2019;**20**(14):1-11

[95] Mutter J, Naumann J, Sadaghiani C, Schneider R, Walach H. Alzheimer disease: Mercury as pathogenetic factor and apolipoprotein E as a moderator. Neuroendocrinology Letters. 2004;**25**(5):331-339

[96] Leon-Mejia G, Quintana M, Debastiani R, Dias J, Espitia-Perez L, Hartmann A, et al. Genetic damage in coal miners evaluated by buccal micronucleus cytome assay. Ecotoxicology and Environmental Safety. 2014;**107**:133-139

[97] Carre J, Gatimel N, Moreau J, Parinaud J, Leandri R. Does air pollution play a role in infertility? A systematic review. Environmental Health. 2017;**16**(1):82

[98] Mahalingaiah S, Hart JE, Laden F, Farland LV, Hewlett MM, Chavarro J, et al. Adult air pollution exposure and risk of infertility in the nurses' Health Study II. Human Reproduction. 2016;**31**(3):638-647

[99] Merklinger-Gruchala A, Kapiszewska M. Association between PM10 air pollution and birth weight after full-term pregnancy in Krakow city 1995-2009 –trimester specificity. Annals of Agricultural and Environmental Medicine. 2015;**22**(2):265-270

[100] Arrhenius S. Worlds in the Making. London and New York: Harper and Brothers Publishers; 1907. p. 54

[101] Houghton J. Global Warming: The Complete Briefing. 4th ed. UK: Cambridge University Press; 2009

[102] Boot-Handford ME, Abanades JC, Anthony EJ, Blunt MJ, Brandani S, Mac Dowell N, et al. Carbon capture and storage update. Energy & Environmental Science. 2014;**7**:130-189

[103] World Coal Institute. Coal: Meeting Global Challenges. London: WCI; 2006

[104] NETL. Baseline Technical and Economic Assessment of a Commercial Scale Fischer-Tropsch Liquids Facility, Final Report, DOE/NETL-2007/1260. Pittsburg, PA: US Department of Energy NETL; 2007

[105] International Energy Agency. Clean energy technologies. 2020. Available at: https://www.iea.org/reports/clean-coal-technologies

[106] Cohan D. Coal Is Not Competitive with Cheaper Alternatives, and the industry's True Costs Are Even Higher. Oakland, CA: GreenBiz, GreenBiz Group; 2018 Article dated 8 November 2018. https://www.greenbiz.com/article/coal-notcompetitive-cheaper-alternatives-and-industrys-true-costs-are-even-higher

[107] Stallard E. Breakthrough in nuclear fusion energy announced. BBC News Climate and Science; 2022. Available from: https://www.bbc.com/news/science-environment-63957085 [Accessed: December 15, 2022]

Chapter 2

Perspective Chapter: The Japanese Coal Mining Industry Reconsidered – From Mechanized Longwall Mining to Carbon Dioxide Capture and Storage

Tomoki Shimanishi, Taku Shimizu, Naoko Shimazaki, Ken Takahashi and Shigeo Nakajima

Abstract

This chapter investigates the role of the Japanese coal mining industry in global coal mining development. In the twenty-first century, the Japanese coal mining industry was a marginal contributor to global production, with an annual domestic production of only 750,000 tons. However, as explained below, Japan has contributed to clean coal technologies and coal mining. The combination of chock shields and a double-ended ranging drum shearer is one of the major mechanized longwall mining systems worldwide. It was developed by the Taiheiyo Coal Mining Company in Japan in the mid-1960s and subsequently spread among major coal-producing countries. Japanese coal production systems and clean coal technologies have been transferred to Asian countries since the late-1980s. Currently, carbon dioxide capture and storage system in a mine under the seabed is being implemented by the Kushiro Coal Mine Company which succeeded the mine of Taiheiyo. In addition, Japan has developed technologies for coal-fired power plants to burn coal more efficiently and contains the world's most efficient ultra-supercritical coal-fired power plants. Furthermore, the world's largest Integrated Coal Gasification Combined Cycle power plants started operating to reconstruct Fukushima and reduce carbon emissions.

Keywords: Japanese coal mining industry, longwall system, technology transfer, coal gasification combined cycle, carbon dioxide capture and storage

1. Introduction

This chapter clarifies the role of the Japanese coal mining industry in the global development of coal mining in terms of technology and economic history. In addition, it investigates the role of Japanese technologies in utilizing coal in the trend of global decarbonization.

According to the International Energy Agency (IEA), the total amount of coal consumption in 2024 will increase by 0.3% compared to 2021, driven by increased coal consumption in China, India, Southeast Asia, Russia, and Africa, although developed countries such as G7 are tackling the issue of decarbonization [1]. However, decarbonized ironmaking technologies such as hydrogen ironmaking have not yet been operated commercially [1]. The supply of Liquefied Natural Gas from Russia became unstable following Russia's invasion of Ukraine. Therefore, the amount of global coal consumption will possibly increase more in the near-to medium-term.

In particular, the IEA estimates that the amount of coal consumption in the Asia Pacific will increase by 1.7% and account for 80% of global coal consumption, while Japan's coal consumption will decrease by 3.6% in 2024 compared to 2021. Coal production in China, India, and Indonesia accounts for 70% of global production. Including Australia, this increases to 75% of the global production [1], making the Asia-Pacific region the world's largest producer and consumer of coal.

In China, the world's largest coal-producing country, 87% of the total coal production comes from underground mines [2]. In contrast, open-cut coal mines are dominant in India, Indonesia, and Australia [1, 3]. However, coal mines have started mining underground in these countries because of a decrease in the amount of surface coal reserves, an increase in the amount of stripped soil, and various regulations for protecting the natural environment [3]. As a forerunner from open-cut mining to underground mining, the Vietnamese coal mining industry produces 73% of its coal from underground mines and will close all open-cut mines by 2030 [4]. The coal mining industry in the Asia-Pacific region has an increasing need for underground mining.

The Japanese coal mining industry produced a mere 60 million tons per annum, even in its flourishing period, which quickly declined after the late 1960s. Despite this, there exists a relationship between the above-mentioned situation in the global coal market and the Japanese coal mining/consuming industry. The global standard method of longwall mining, the combination of a double-ended ranging drum shearer (hereafter shearer) with chock shields, is based on the "Shield support and Drum shearer (SD) system" established by Japan's Taiheiyo Coalmining Company (Taiheiyo) in Kushiro City, Hokkaido, in the 1960s–1970s. Taiheiyo was a pioneer in longwall mining mechanization. At present, Kushiro Coal Mine Company (KCM) is part of a technology transfer project by the Japanese government.[1] Engineers who operated the SD system at Taiheiyo's colliery transferred knowledge on longwall mining and mine safety to other Asian countries, such as China, Vietnam, and Indonesia.

Japan imports more than 100 million tons of coal per annum and is the world's third largest coal importer after China and India. The Japanese electric power industry and heavy electrical machinery manufacturers have developed coal-fired power-generating technologies, introducing larger power-generating capacity, increasing thermal efficiency, and research and development of Integrated Gasification Combined Cycle (IGCC) power generation. Currently, the Japanese electric power industry operates many ultra-supercritical (USC) coal-fired power plants that have the world's highest gross thermal efficiency. These coal-fired power plant technologies are also exported to Asian countries to reduce carbon emissions, although it is becoming difficult to export these technologies due to the expansion of decarbonization in the world.

---

[1] Taiheiyo stopped operating mines in 2002. Thereafter, KCM was established to inherit the operation of the mine staked by the local businesses in Kushiro City.

Former Prime Minister Yoshihide Suga declared the goal of "2050 Carbon Neutrality," which demonstrated that the Japanese government will aim to decarbonize the economy in step with G7 countries. However, this does not mean that Japan has to abandon the technologies for coal mining and utilization that it has accumulated. It is an important duty of Japan, one of the world's largest coal consumers, to correctly record the experience of developing coal-related technologies and transfer them to countries that need them as it takes a long time to achieve energy transition [5].

Many studies have focused on the Japanese coal mining industry after the Second World War, including policy studies of industrial policies for rationalizing the coal mining industry [6, 7] and historical studies of workers and their communities [8–11]. These studies clarified the process of industrial decline by focusing on the government, companies, workers, unions, and communities. However, research on coal-related technologies from the perspective of social science has only recently begun [12–15]. This chapter contributes to improving the understanding of coal-related technologies in Japan.

2. Brief history of the Japanese Coal Mining Industry

According to an investigation in 2008, Japan's proven and recoverable coal reserves stand at 4.8 billion tons and 352 million tons, respectively [16]. The most unique characteristic of coal reserves is the existence of subsea coalfields, which comprise 18% of the total coal reserves. Annual production from reserves has been almost 100% of the total production since the late 1990s, while the corresponding number was 12% in 1958 [17, 18]. This indicates that production from subsea coalfields is essential for the Japanese coal mining industry.

As shown in **Figure 1**, the modern Japanese coal mining industry started to develop in the late 19th century and declined after 1961, when it produced approximately 55

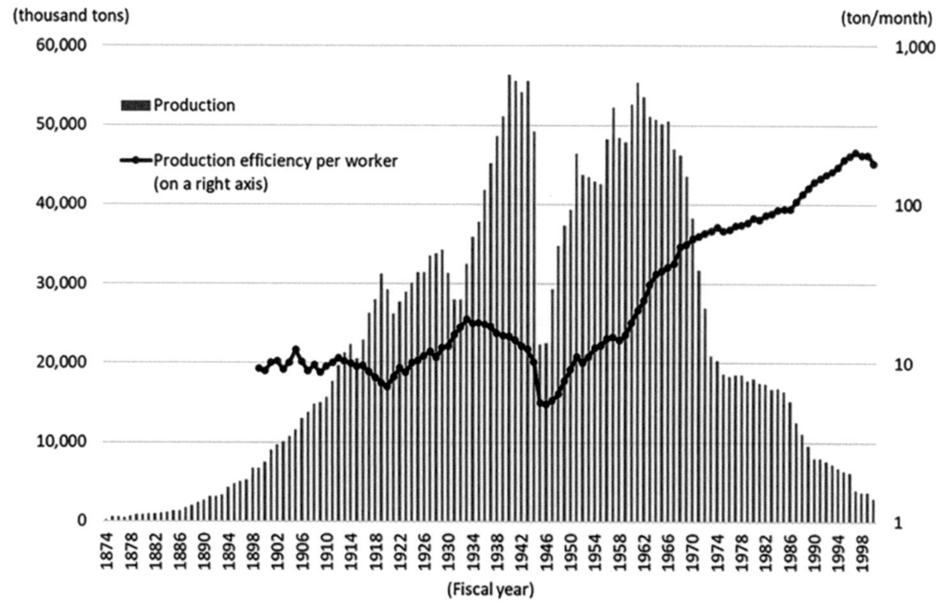

Figure 1.
Coal production and production efficiency in Japan, 1874–2000. Source: [18].

million tons. Today, some coal mines produce a mere 750,000 tons per annum in 2020 [19]. On the other hand, production efficiency increased rapidly despite the decline in industry. The monthly production efficiency per worker was 200 tons in the 1990s, compared to only 5 tons in the late 1940s. This indicates the rapid mechanization of production as well as a rapid decrease in the number of workers.

The Japanese government launched a set of policies for the rationalization of the coal mining industry in 1963 to deal with the industry's decline caused by the energy revolution. Financial support measures, such as concessional loans, subsidies, and leases of machinery, were implemented to promote investment by coal companies in facilities and machinery. This financial support was provided by the government, the most notable of which was the import petroleum tariff after 1967. Financial support from the government continued until FY 2001, when the total amount reached staggering proportions [18]. For example, concessional loans alone amounted to approximately 330 billion yen, that is, approximately $2.4 billion.

Furthermore, Sekitan Gijutsu Kenkyuusho (Coal Mining Research Center) was established in 1960 through joint investment by the government and coal companies. Engineers from both coal and machinery companies collaborated for experiments and research projects to enable innovation in coal production technologies [15]. Thus, the coal mining industry has created institutional and financial support to spur technological innovation.

However, the coal mining industry has been forced to introduce basic technologies from abroad. Mining engineers not only collected mining journals from Europe and the United States but also sent overseas inspection groups to study advanced coal production technologies [20]. In addition, mining engineers have studied the coal production technologies of the USSR because many Japanese coal mines excavated coal using inclined shafts, such as used in mines in the USSR, in contrast to the vertical shafts in continental Europe [21]. Following these experiences, various types of mining machinery were imported. Thus, the coal mining industry promoted the mechanization of coal excavation. One of the most notable successful examples is the mechanization of longwall mining, which Taiheiyo developed as the "SD system" in the late 1960s. Taiheiyo operated a subsea mine in Hokkaido, on the northern island of Japan. In the next section, we will investigate Taiheiyo's technological development history.

3. Struggle to mechanize longwall mining: Development of the SD system

In the 1960s, Taiheiyo made an effort toward the practical application of powered support. In May 1960, Ferromatik rotating powered supports, which were frame-type powered supports, were introduced as Taiheiyo's first powered support. However, because of the small contact area of the base of the powered support, insufficient thrust of the shifter, insufficient strength of each member, and other factors, Taiheiyo was unable to obtain the desired results [22]. Next, the company introduced an IU-shaped, frame-type, six-leg powered support (MKSP-LIU model, FIU model) manufactured by Mitsui Miike Machinery (Miike Machinery). A low seam with a coal height of approximately 1 meter was selected for the machine to conduct operations. The coal-cutting machine adopted a tandem-type coal plow, where the machine height changed according to the height of the coal face, and the machine began full-scale operation in October 1963 [23]. This low-seam mining achieved good coal output after partial improvement of the IU frame (from the LIU model to the FIU model) [24]. However, many occupational accidents occurred because of the narrow space of the low seam [25].

In parallel with low-seam mining, Taiheiyo aimed for the practical use of powered supports in thick seams. When designing the powered support, it is necessary to estimate the amount of roof lowering when using the powered support and determine the stroke amount required for the hydraulic prop to avoid the complete retraction of the powered support on the coal face. Therefore, experiments that simulated repeated lowering and sequential lowering, which are shoring movements unique to powered support, were conducted using hydraulic props and link bars, and the roof subsidence behavior when using the powered support was clarified from the measurement results [26, 27]. Based on their findings, a UU-shaped, frame-type, 12-leg powered support (MKSP-FUU model) was developed by Miike Machinery, which had two hydraulic props in the rear row. UU frames (UU) were introduced for simultaneous multi-pass longwall mining in combination with a coal plow and underground tests began in May 1964. Various modifications, such as changes to the method for connecting the hydraulic prop and base and the shortening of the base length, were made to the UU based on the findings obtained through the tests [28–30].

In February 1965, Numajiri No. 2 Longwall began operations using the improved UU. However, the expected efficiency improvements were not achieved. In the UU, hydraulic props with corresponding relationships were connected to each other at the front, back, left, and right. This resulted in the frequent occurrence of frame inclining owing to the relative horizontal movement of the roof and floor, as well as hydraulic prop damage owing to eccentric loading. Additionally, there were many structural problems, such as the need to correct the direction of the powered support, forepoling, and inflow of rock debris from the gob area. As a result, mechanized mining of this thick seam failed [22, 31, 32]. In the section where an attempt was made at the practical application of UU, from the second half of 1965, the lower-seam longwall for simultaneous multi-pass longwall mining was switched from UU frames to single hydraulic props and link bars.

Efforts toward the practical application of powered support continued and surplus UU frames that were lifted to the surface were dismantled. After dismantling, the materials were used to develop an in-house THY chock support, which is a chock-type powered support that is connected to a conveyor. THY chocks were tested in combination with UU frames. The linearity of the powered support was improved by adopting a conveyor connection. Additionally, to improve frame inclining, an adjusting jack was made and installed using scrap material from a single hydraulic prop and a simple shield plate was made and attached to prevent the inflow of rock debris from the roof and gob area [31, 32].

However, damage to the hydraulic prop owing to the horizontal force of the coal face roof was structurally unavoidable with the frame-type powered support and chock support. At that time, the only powered support with a structure that could withstand horizontal forces was the OMKT shield support (OMKT) developed in the USSR. As shown in **Figure 2**, the OMKT was a type of powered support called a shield support, which was covered with steel plates on the top and rear. The OMKT used a pin-hinge structure to hold both ends of the hydraulic prop and no horizontal force or bending moment was applied to the hydraulic prop. In December 1965, Miike Machinery formed a technical collaboration with the USSR's export corporation for the OMKT and began preparations for licensed production [33]. Taiheiyo received 20 sets from Miike Machinery for trial use [32].

In addition, a shearer was adopted for the coal-cutting machine combined with the OMKT. Equipping both sides of the machine body with ranging arms with

Figure 2.
The OMKT shield support. Source: This figure is owned by the Kushiro Board of Education, Hokkaido, Japan.

cutting drums made it possible to cut both ends of the coal face. The double-ended ranging drum shearer was domestically produced by Miike Machinery in 1966 [34]. The MCLE200-DR8090 model manufactured by Miike Machinery was delivered to Taiheiyo. A coal planer with a modified coal plow that can load cut coal and cut the bottom wall was installed to supplement the cutting and loading by the shearer [35].

In addition to changing the cutting machine and powered support, the longwall coal mining method was fundamentally reconsidered. The length of the coal face was set extremely short at 35 meters to keep the load on the roof at the center of the coal face small. Furthermore, shortening the round-trip time of the shearer and accelerating the progress of the coal face enabled the advancement of the powered support before the deterioration of the roof, and it was thought that the complete retraction of the powered support could be prevented [35].

To handle this rapid progress, the stable was abolished because of the difficulty of mechanizing it and the requirement of manual labor. First, the drive unit, which was conventionally installed on both stables on the head-gate and tail-gate sides, was integrated into the head-gate side, and the stable on the tail-gate side was abolished. The return end of the armored face conveyor (AFC) on the tail-gate side and the sprocket of the coal planer were coaxial to avoid interference between the cutting drum and AFC, which allowed the drum to enter the end of tail-gate side and complete the cutting of the longwall. Next, all the drive parts on the head-gate side were pulled out to the head-gate. This was made possible by adopting a separator system at the entrance from the AFC to the head-gate conveyor. As a result, the stable on the head-gate side was also abolished and became a non-stable system. Additionally, a snaking panzer conveyor was developed to respond to the rapid relocation of facilities such as the head-gate conveyor. This conveyor was installed parallel to the normal straight panzer conveyor that was installed at the head-gate and, by overlapping the snaking tip, it became possible to shorten or lengthen the head-gate conveyor without interfering with the operation on the coal face [35]. These head-gate facilities later led to the development of the stage loader.

This coal mining method was called "SD system," which was an acronym for the main constituent machines: the shield support and drum shearer. In April 1967, the test coal face Higashi-Masuura Left No. 1 Level Rise No. 4 Longwall was set up and the first SD system began operation. Despite facing difficulties, such as mechanical failure and roof fall, the equipment was improved and the number of props built in the OMKT was increased from the original one to two to increase the supporting capacity. Progress was generally favorable after the introduction of the reinforced OMKT in Longwall no. 10 [35].

However, the OMKT, which was originally developed for shallow areas, could not overcome the lack of structural support. Therefore, Taiheiyo and Miike Machinery, learning from the OMKT-MK shield support, which had a lemniscate linkage at the connection between the caving shield and base, developed a new chock shield that had a long main canopy, four hydraulic props that supported it, and a lemniscate linkage at the connection between the base and the caving shield. The adoption of the lemniscate linkage enabled nearly vertical retraction of the entire canopy while maintaining a fixed distance between the canopy tip and faceline. This four-leg chock shield with a lemniscate linkage was called the "SMK chock shield support (SMK)" [33, 36] (see **Figure 3**).

In December 1968, 10 sets of SMK were introduced to Numajiri West No. 23 Level No. 1 Main-seam Longwall, along with single hydraulic props and link bars, and tests were conducted. In November 1969, operation began at the Minami-Masuura No. 3 Level No. 3 Longwall, where the SMKs were installed across the entire coal face [37]. The coal output was good after the introduction of SMK and the supporting capacity

Figure 3.
The SMK chock shield support. Source: This figure is owned by the Kushiro Board of Education, Hokkaido, Japan.

of the powered support was completely resolved. The supporting capacity of the powered support increased from 80 tons for the first monopod-type OMKT to 160 tons for the reinforced OMKT. Following the development of SMK, improvements and strengthening were conducted as the length and depth of the coal face increased. The TY-2 model introduced in 1968 had a capacity of 320 tons, the TY-3 model introduced in 1971 had a capacity of 400 tons, and the NT-1 model introduced in 1988 had a capacity of 600 tons. The length of the coal face gradually expanded from 35 to 100 m in the 1970s, peaking at 250 meters in the 1990s [38, 39].

The four-leg chock shield with lemniscate linkage was a well-balanced powered support that could withstand the complex and difficult underground natural conditions of Japanese coal mines. The development of this four-leg chock shield with a lemniscate linkage was a breakthrough not only in Japan but also in coal mining technology worldwide. This technical information spread worldwide through a licensing agreement with the West German company Westfalia in 1971, published in the West German magazine *Glückauf* Vol. 109 No. 14, exports of the longwall system to Australia in 1975, and international sales by the manufacturer Miike Machinery [32, 33]. It can be said that today's standard longwall system was established through this series of technological developments. This establishment not only brought about a dramatic improvement in coal production efficiency but also made it possible to protect miners from the inflow of rock debris from the gob area and roof falls, which contributed to occupational safety.

4. Overseas transfer of coal production technologies

4.1 Export of SD system and related technologies

4.1.1 Exports of SD system to Australia and other countries

In 1975, the SD system developed in Taiheiyo was exported to the South Bulli Coal Mine (South Bulli) of the Bellambi Coal Company in Australia. The powered support that constituted the longwall system was a four-leg chock shield with a lemniscate linkage made by Miike Machinery, with a yield load of 480 tons and support width of 1.25 m. A DR8394 model manufactured by Miike Machinery, which had a motor output of 300 kW, was adopted for the double-ended ranging drum shearer [34].

The South Bulli mine introduced British and West German chock supports since it began longwall mining in the mid-1960s but both failed because of the lack of support. The roof of the 2.4–2.7-m height Bulli seam, which is the main coal seam for operation in the mine, is Coal Cliff Sandstone, which is a superhard rock, and it is difficult to handle the roof pressure derived from it. In April 1973, Taiheiyo was asked to consider adopting an SD system to break this deadlock and a letter of intent was issued in March 1974. Taiheiyo Kouhatsu, a parent company of Taiheiyo, then established Taiheiyo Engineering Inc. (TEI) in July 1973 to establish an export system.

The supporting capacity of the powered support was given the highest priority when designing the longwall system. The measurement results communicated by the chief engineer at Bellambi indicated a necessary meter run for the face of 320 tons/m for the Bulli seam and a powered support with a yield load of 480 tons was designed based on this result. It was determined that the load capacity of each hydraulic prop must be increased to 120 tons to meet the performance requirements. Sumitomo Metal Industries developed a new large-diameter drawn steel pipe to satisfy these requirements.

Following test chock shield load tests at a research institute for coal mining in West Germany in November 1974, the entire system was tested outside the mine in June 1975, with operations beginning at the coal mining site in October 1975. When operations began, there were problems with the operation of the machine owing to the large frictional resistance caused by the weight of the shearer, as well as problems with the operation of the coal planer for loading coal. Improvements were made through on-site creativity and the issues were resolved. Importantly, the supporting capacity of the powered support, which could have fatal consequences in the event of a defect, withstood the first weighting of the starting section and functioned as designed. This was revolutionary because it succeeded in designing a longwall system based on measurements and knowledge from mining theory. The success of longwall mining also contributed to company management. Sydney Morning Herald reported that the net profit for the first half of 1976–1977 was 447% of that of the previous period. Subsequently, the longwall system continued to be used without any maintenance outside the mine until its planned end in 1986 [32, 34].

As shown in **Table 1**, the SD system and shearers produced in Japan were exported to other parts of Asia, Australia, and the United States [34, 41]. Less than 10 years after the SD system was introduced in Japan, it became a global standard for longwall mining. However, machinery manufacturers in Europe and the United States started increasing exports of longwall systems similar to the SD system after the 1980s because they succeeded in producing a power support akin to the Japanese design [34].

4.1.2 Participation in the project led by the United States Bureau of Mines

The TEI also participated in a project by the United States Bureau of Mines (USBM). In 1974, the USBM put out a public offering for a new technology for mining the thick seams found in the western United States to replace the widely used room coal mining in the United States. Consolidation Coal, Foster–Miller, and TEI were jointly awarded the order. Foster–Miller approached the TEI in the summer of 1974 and joined the project. The outline of the proposal was to conduct non-simultaneous multi-slice mining in advance of the upper layer based on the SD system developed by Taiheiyo.

Year	Destination	Unit
1975	Australia	1 set of longwall system and 1 shearer
1978–1979	China	1 set of longwall system
1980–?	North Korea	3 sets of longwall system
1982	Australia	1 set of longwall system
1985	Australia	1 shearer
1985–?	United States	9 shearers
1986	Australia	2 shearers
1991	Australia	1 set of longwall system

Notes: This table shows only the export products produced by Miike machinery from the 1970s to the 1990s.

Table 1.
Exports of the SD system and shearers from Japan. Source: [34, 40].

The measurement and analysis results of various rock behaviors accumulated at Taiheiyo's colliery were used as mining evidence for this proposal. First, the surface subsidence mechanism caused by the breaking of the upper rock strata and compression of fallen rock debris in the gob area during thick-seam mining was clarified based on actual measurement data on the changes over time in surface subsidence in the cases of single-seam mining and multi-seam mining at Taiheiyo's colliery. Additionally, the solidification state of fallen rock debris in the gob area of the upper layer, which becomes an artificial roof when mining the lower layer in multi-slice mining, was clarified from actual measurement data on the rock solidification of the main seam at Taiheiyo's colliery. Its low porosity indicated that it could be handled by chock shields. Furthermore, the mechanical and technical evidence included references to the practical application process for powered support at Taiheiyo's colliery, with the experience from use in the field and the improvement process shown by the introduction of the frame-type powered support made by Miike Machinery (i.e., UU frame) to the in-house developed chock support (i.e., THY chock support), OMKT, and four-leg chock shield with a lemniscate linkage.

This mining and practical knowledge were highly regarded and led to the order of this project. In April 1975, a proposal entitled "the Technical Proposal for the Demonstration of a Thick Seam Mining System" was created and submitted to the USBM [32].

4.2 Technology transfer from Japan to Asian countries

The Japan Technical Co-operation Center for Coal Resources Development (JATEC), which was established by a joint investment by many coal mining companies to investigate and develop overseas coal resources in 1990, started receiving technical trainees from the Ombilin Coal Mine in Indonesia [42]. Simultaneously, the Ministry of International Trade and Industry (MITI) devised a project for mine safety improvement and increased production in the Asian coal mining industry through technology transfer from Japan. JATEC undertook the technical training program component of the project from MITI and was entrusted with implementing the program with coal mining companies, including Taiheiyo. After this project was successfully completed, a new and enlarged technology transfer project was implemented by the New Energy Development Organization (NEDO) in the late 1990s [13, 14]. Finally, in FY 2002, the Ministry of Economy, Trade and Industry (METI), that is former MITI, devised a five-year project, where NEDO maintained the operation of domestic coal mines and served as a technology transfer hub to Asian coal mining industries [18].[2] KCM was entrusted with the operation of the project from NEDO because it inherited the coal mine that Taiheiyo had stopped operating at the end of FY 2001.

As shown in **Table 2**, the technology transfer program included the reception of technical trainees and dispatch of mining engineers. Between FY 2002 and FY 2013, the program included 2735 technical trainees from Vietnam, China, and Indonesia, 3680 Japanese mining engineers, and 88,074 trainees who completed training in the three above-mentioned countries. In terms of the number of people involved, the program in Vietnam was the largest.

To demonstrate the program details, the case of KCM, which has implemented the program since FY 2002, will be outlined here. KCM received 367 trainees in five

[2] After FY 2012, the Japan Oil, Gas and Metals National Corporation (JOGMEC) inherited the project from NEDO.

Recipient country	China	Vietnam	Indonesia	Total
Received trainee	953	1344	438	2735
Dispatched engineer	467	2553	1440	3680
Trainee at home	19,249	54,000	14,825	88,074

Unit: Person.
Notes: The reception of trainees from Indonesia was conducted from FY 2002 to FY 2009. The number of dispatched engineers and trainees at home is not the actual count but the total count.

Table 2.
Results of the coal technology transfer program, FY 2002–FY 2013. Source: [43, 44].

training courses spanning 10–24 weeks from FY 2002 to FY 2003. The curriculum was as follows: on-the-job underground work using mining machines, classroom lectures on the Japanese language and business management, and demonstrations on mining methods using a miniature model of a mine and computer technologies [12]. Mining engineers dispatched from the KCM conducted practical training and classroom lectures at overseas mines for 3 months. Although the transfer of production technologies was partial, the technology transfer project won high praise from recipient countries because of the successful transfer of knowledge on mine safety and mining operations [12].

In recipient countries, the technology transfer program contributed to the production rationalization of the coal mining industry; for example, an increase in production efficiency, a decrease in the occurrence rate of mine accidents, and heightened awareness of technology improvement and mine safety among ex-trainees [45]. Due to this, the technology transfer project continues to this day, despite originally being planned to last for 5 years.

KCM needs to continue implementing the technology transfer program because the program contract fee is one of its main sources of revenue [42]. If the project was abolished, the KCM could not continue operating its mines. This means that the Asian coal mining industry would lose the chance to learn about the production technologies that the Japanese coal mining industry has fostered over its history. As mentioned in Section 1, there is an urgent need for underground mine safety transfer because coal mines have increasingly started underground operation in Asian countries because of a decrease in the amount of surface coal reserves, increase in the amount of stripped soil, and various regulations for protecting the natural environment [3]. Therefore, the Japanese government and METI have to continue the technology transfer project to maintain KCM's operation and make Asian coal mines safer and more efficient.

5. Coal as clean energy: USC, IGCC, CCS, and Carbon Capture, Usage and Storage (CCUS)

5.1 Coal-fired power plants

As long as coal-fired power generation is still used in China and worldwide, improving its efficiency and applying clean coal technologies remains an urgent issue. Japan has accumulated technologies that can solve these problems in conventional coal-fired and IGCC power generation.

The late 1950s saw the start of the shift from coal to oil as the main energy source in Japan, or the energy revolution. During this period, the Japanese electric power industry promoted a power-supply approach that involved a shift from reliance on hydropower to thermal power. The electric power industry continued to use domestic coal based on government and coal industry requests, while increasing the number of heavy oil-fired power plants that used inexpensive Middle Eastern heavy oil as fuel. Meanwhile, the electric power industry, in cooperation with Mitsubishi Heavy Industries, Hitachi, IHI, and others, worked on domestic production and the improvement of imported technologies. By the late 1960s, the Tokyo Electric Power Company (TEPCO) started operating a 600-MW supercritical pressure power plant with a net efficiency of 40.3%, catching up with the plants in operation in Europe and the United States [46]. In the early 1980s, J-POWER began operating a 500-MW supercritical coal-fired power plant. Furthermore, in the 1990s, J-POWER successfully operated a 1000-MW USC coal-fired power plant. Currently, Japan's coal-fired power generation technology is at the highest level in terms of power generation efficiency worldwide, with the net efficiency of the country's most efficient power plants reaching 48%, which has contributed to reducing coal consumption and CO_2 emissions [47, 48].

Simultaneously, the scale of conventional coal-fired power generation has progressed and research and development into IGCC has also progressed. The description below is based on cited references and an interview with Yoshihiko Horie, Plant Chief Superintendent of Nakoso IGCC Power LLC, on 17 November 2022. IGCC is a technology that gasifies coal and generates electricity with a gas turbine and steam turbine that uses exhaust heat, with research and development beginning in Europe, the United States, and Japan in the 1970s. Japan's unique air-blown IGCC technology began in 2007 with a 250-MW demonstration unit at the Nakoso Power Plant of Joban Joint Thermal Power in Fukushima Prefecture. The demonstration unit was successful, achieving a high net efficiency of 42.9% and began commercial operation in 2013 [49].

In 2015, the Japanese government announced a policy aimed at accelerating recovery from the 2011 Great East Japan Earthquake and Fukushima Daiichi Nuclear Power Plant Accident. In response to this policy, TEPCO, Joban Joint Thermal Power, and the Mitsubishi Group roughly doubled the scale of the demonstration unit at Nakoso and launched the Fukushima Recovery Power Project to build and operate the world's largest and most efficient IGCC thermal power plant in Fukushima, establishing Nakoso IGCC Power LLC and Hirono IGCC Power LLC in 2016. In this chapter, the case of the Nakoso IGCC Power is discussed. The location selected for the new power plant in Nakoso was agricultural land that suffered salt damage due to the tsunami caused by the earthquake and the project created local employment for those affected. Engineers at Nakoso faced various technical problems that were associated with the scaling up of the unit but the engineers proceeded with work to begin power generation in 2020 while resolving these issues and started commercial operation in April 2021. As of 2022, the Nakoso IGCC power plant had an output of 525 MW and net efficiency of 48%, making it the world's largest and most efficient coal gasifier on a per unit basis.

IGCC is a power generation technology with a lower environmental impact than conventional coal-fired power generation methods. The first benefit is the reduction in greenhouse gas emissions. The CO_2 emissions of IGCC power plants are 15% lower than those of conventional coal-fired power plants. The second benefit is the expansion of the type of coal that can be used. Conventional coal-fired power generation uses coal with a high ash melting point, which is less likely to generate clinkers. In contrast, IGCC power generation can use coal with a low ash melting point. The third benefit is that coal ash becomes slag-like, which not only halves the volume needed but also allows the material to be reused as cement or pavement material.

In the 2020s, Japan's main coal supply sources were Australia and Indonesia, and conventional thermal power generation uses coal from these sources, whereas the Nakoso IGCC power plant uses American coal. Combining IGCC power generation with conventional coal-fired power generation enables the reduction of the environmental impact of coal combustion and efficient use of resources. The COVID-19 pandemic and the effects of "decarbonization" in developed countries have resulted in the stagnation of IGCC technology exports from Japan to overseas. However, since a stable supply off fossil energy is expected to become unstable, the value of IGCC technology should be revised upward. Additionally, NEDO, J-POWER, and Chugoku Electric Power Company have been conducting Integrated Coal Gasification Fuel Cell Combined Cycle demonstration tests by applying the IGCC technology (Osaki CoolGen Project) in Hiroshima in 2022 [50, 51].

Despite the decline in the domestic coal mining industry, one of the reasons for the advancement of coal-utilization technologies in Japan was the import of steam coal, which was promoted as a countermeasure to the soaring oil prices seen after the oil crisis. At the beginning of the 1970s, unlike the coking coal trade, where long-term contracts were the norm, the steam-coal trade commonly involved spot contracts and a large-scale global market based on long-term contracts had not yet formed. Under these circumstances, in 1974, Australia, which had previously banned the export of energy resources such as uranium, petroleum, and steam coal, lifted its ban on the export of steam coal [52]. This was achieved through negotiations between the Australian government and J-POWER, which was planning to build a large-capacity coal-fired power plant utilizing imported steam coal, and a summit meeting between the Japanese Prime Minister (Kakuei Tanaka) and Australian Prime Minister (Gough Whitlam) in November 1974 [53]. Subsequently, Japan increased its imports of steam coal, including from Australia, America, Canada, China, and Indonesia. Japan also contributed to the expansion and globalization of the steam-coal trade market.

5.2 Challenging coal production and utilization with CCUS at KCM

The description below is based on an interview with Hiroyuki Matsumoto, Senior Managing Director of KCM, on 22 March 2022. The KCM is working on a resource recycling project that integrates the operation of an underground coal mine and thermal power plant. This is based on the Kushiro Resources and Energy Ecopark Concept, which was formulated by the company in 2011. The concept was to organically link the company's coal production, training, and environment-related businesses and to establish a resource and energy-related research and education center, the core of which is the construction of a new thermal power plant. In Kushiro City, the local government also supported the power plant construction plan because it was informed that there was a risk in the surrounding area due to the uneven distribution of the power supply as a result of the absence of a core power plant.

The Kushiro thermal power plant, which started operation in 2020, has a small output of 112,000 kW and uses a circulating fluidized bed (CFB) boiler to conduct mixed combustion of Kushiro coal and overseas biomass. The biggest advantages of adopting a CFB boiler instead of pulverized coal firing are that there is no need for fuel pretreatment, the mixed combustion of biomass and waste is easy, and the fuel can be changed based on the circumstances.

Additionally, KCM has been entrusted with the operation of the Kushiro Wide-Area Federation Incineration Plant. It dispatched engineers and operators and has had a track record of the stable operation of a fluidized-bed gasification melting furnace and small-scale power generation plant there. On the other hand, the calories from coal and biomass are almost constant in the new power plant. Thus, boiler operation is less difficult than general waste incineration, which involves large calorie fluctuations. Having the manpower of experienced boiler operators was a stepping-stone in the thermal power generation plan.

A CFB boiler uses less cooling water than a pulverized coal-fired boiler. As a result, although a supply line from the water supply is secured, most of the aforementioned demand can be covered by the underground water of the coal mine. Furthermore, the thermal wastewater discharged from thermal power generation is returned to the coal mine and used as coal-preparation water. Even as a coal preparation plant, this has the advantage of preventing the freezing of the coal-preparation water in winter. Fly ash discharged from thermal power generation is also supplied to coal mines and is used as a filling material in underground mining pits.

In underground coal mines, the company changed its production system to achieve integrated operation with thermal power plants. Because the production scale was reduced from 500,000 tons per year to 300,000 tons per year for local production and consumption, the one-coal-face system of longwall mining was abolished and switched to the four-coal-face system of room and pillar mining. With the top priority of a stable supply of coal to local thermal power plants, the company shifted from a centralized production system to a risk-diversified production system. The company called this the New Coal Production System.

A unique method was adopted for room and pillar mining as part of a Resource Recycling Project. In normal room and pillar mining, mining is conducted in a grid pattern while leaving security coal pillars and mining traces are not filled. However, the room and pillar mining conducted by KCM differs from the usual mining method in that it extracts coal by digging branches into the coal bed and fully filling each branch with fly ash. As will be described later, technological development is being conducted to connect this full filling to Carbon Dioxide Capture and Storage and Carbon Capture, Usage and Storage (CCUS).

The KCM is actively conducting joint research with universities, academic societies, and companies. Although the number of underground mines in operation in Japan has decreased, the mine has served as a valuable research site for resource engineering, geology, and rock mechanics. For example, as part of a resource recycling project, KCM is working with the Osaka Gas Company to develop concentration technology for underground methane gas, which is used as fuel for its in-house boilers. Currently, joint research is being conducted on CCUS. This involves mixing carbon dioxide with fly ash, which is used as a filling material for room and pillar mining. This carbon dioxide reacts with the Ca and Mg in the fly ash and improves the strength of the filling material through carbonate mineralization. A demonstration test was conducted and favorable results were obtained. As this is currently in the demonstration test stage, carbon dioxide is externally procured; however, following the completion of the demonstration test, the company is looking to install carbon dioxide separation and capture equipment at the thermal power plant.

As shown above, KCM has been able to utilize its technological base and human resources to operate its underground coal mine and thermal power plant in an integrated manner and is positioned as a base for various research and development efforts. KCM is trying to find a way forward as an active coal mine and thermal power plant amidst the trend of decarbonization by continuing to present added value in the form of research and education that goes beyond the production and use of coal.

6. Conclusion

The Japanese coal mining and electric power industries have contributed to the development of coal-fired power generation and underground coal mining. The following three contributions can be noted:

The mechanization of longwall mining. In the late 1960s, Taiheiyo combined a shearer with powered supports into the "SD system." In this process, the development of a four-leg chock shield with lemniscate linkage was a technical milestone. These technical advancements were possible because Taiheiyo's mining engineers had knowledge of mine engineering and practical techniques and through the close relationship with a mining machine manufacturer, Miike Machinery. The SD system and its related technologies began to be exported abroad after the 1970s.

The transfer of technology to Asian countries. Technology transfer was implemented after the 1990s, when the Japanese coal mining industry was in a continued state of decline. Knowledge of mine safety and mine operations was transferred from Japan to coal mining industries in other Asian countries, which had suffered from decreases in mining efficiency and increases in mining accidents as economic development led to increased coal production. Technology transfer contributed to the modernization and rationalization of coal mines, although it did not provide much financial benefit to the Japanese coal mining industry.

The promotion of thermal efficiency and decarbonization of coal-fired power generation. The Japanese electric power industry succeeded not only in increasing the power-generating capacity of USC power plants but also in the commercial operation of IGCC power plants. In addition, technological developments in the combination of coal-fired power generation with CCUS at an underground coal mine were made by KCM. Large coal-consuming countries, such as China and India, replacing conventional coal-fired power plants with the most advanced ones will accelerate global decarbonization.

Acknowledgements

We would like to thank Hiroyuki Matsumoto, Koyu Sato, Yoshihiko Horie, and Taichi Manome, who cooperated with our interview research.

Funding

This work was supported by JSPS KAKENHI, Grant Numbers 19H00611, 20K01797, and 20K22152.

Author details

Tomoki Shimanishi[1*], Taku Shimizu[2], Naoko Shimazaki[2], Ken Takahashi[3] and Shigeo Nakajima[4]

1 Rikkyo University, Tokyo, Japan

2 Waseda University, Tokyo, Japan

3 No Institution, Tokyo, Japan

4 No Institution, Gunma, Japan

*Address all correspondence to: t.s@rikkyo.ac.jp

References

[1] IEA. Coal 2021: Analysis and forecast to 2024 [Internet]. 2021. Available from: https://www.iea.org/reports/coal-2021 [Accessed: August 26, 2022]

[2] Liu X, Li L, Yang Y. Development status of coal mining in China. Journal of the Southern African Institute of Mining and Metallurgy.2023;**123**(1):19-27. DOI: 10.17159/2411-9717/1506/2023

[3] Sasaoka T, Hamanaka A, Shimada H, Matsui K, Ichinose M. Development of underground coal mine from toe of open-cut highwalls in Indonesia. Journal of MMIJ. 2015;**131**:195-202. (In Japanese). DOI: 10.2473/journalofmmij.131.195

[4] Vietnam News Liner. Double increasing in efficiency of coal mining through introduction new technologies in Quangninh Province [Internet]. 2021. Available from: http://www.n-liner.jp/ [Accessed: August 26, 2022] (In Japanese)

[5] Smil V. Energy and Civilization: A History. Cambridge: MIT Press; 2017. p. 552

[6] Samuels RJ. The Business of the Japanese State: Energy Markets in Comparative and Historical Perspective. Ithaca, NY: Cornell University Press; 1987. p. 358

[7] Hein LE. Fueling Growth: The Energy Revolution and Economic Policy in Postwar Japan. Cambridge, MA: Harvard University Press; 1990. p. 423

[8] Allen M. Undermining Japanese Miracle: Work and Conflict in a Coal-mining Community. Cambridge, New York: Cambridge University Press; 1994. p. 318

[9] Culter S. Managing Decline: Japan's Coal Industry Restructuring and Community Response. Honolulu: University of Hawaii Press; 1999. p. 232

[10] Thomann B. Rationalization, working conditions and living standards in postwar Japanese coalmining communities. Asian Journal of German and European Studies. 2018;**3**:13. DOI: 10.1186/s40856-018-0035-8

[11] Lim T, Shimazaki N, Godo Y, Lim Y. Coal Mining Communities and Gentrification in Japan. Singapore: Palgrave Macmillan; 2019. p. 225. DOI: 10.1007/978-981-13-7220-9

[12] Ishikawa T. Betonamu no Tankou wo Houmon Shite. Kushiro Shishi Kenkyuu. 2006;**5**:95-117. (In Japanese)

[13] Kikkawa T. Shigen Enerugii Seisaku. Tokyo: Keizai Sangyou Chousakai; 2011. p. 459 (In Japanese)

[14] Takeda H. RIcchi Kankyou Hoan Seisaku. Tokyo: Keizai Sangyou Chousakai; 2011. p. 720 (In Japanese)

[15] Shimanishi T. Nihon Sekitan Sangyou no Sengoshi. Tokyo: Keio University Press; 2011. p. 374 (In Japanese)

[16] Japan Coal Energy Center, editor. Sekitan Riyou no Saishin Gijutsu to Tenbou. Tokyo: CMC Books; 2014. p. 300 (In Japanese)

[17] Iki S. Waga Kuni Kaitei Tankou no Mitooshi. In: Agency of Industrial Science and Technology, editor. Sekitan Riyou Gijutsu Kaigi Kaigiroku. Vol. 3. Tokyo: Japan Energy Association; 1959. pp. 16-17 (In Japanese)

[18] Sekitan Seisakushi Hensan Iinkai. Sekitan Seisakushi Shiryouhen. Tokyo:

Japan Coal Energy Center; 2002. p. 405 (In Japanese)

[19] Japan Coal Frontier Organization. COAL Data Book. Tokyo: Japan Coal Frontier Organization; 2022. p. 316 (In Japanese)

[20] Japan Productivity Center, editor. Oushuu ni Okeru Sekitan Kigyou. Tokyo: Japan Productivity Center; 1958. p. 178 (In Japanese)

[21] Soren Tankou Gijutsu Chousa Dan, editor. Soren no Tankou Gijutsu. Tokyo: Nihon Sekitan Kyokai; 1962. p. 150 (In Japanese)

[22] Kishimoto Y, Takasaki M. Taiheiyou Tankou ni Okeru Jisoushiho no Shiyou Keiken ni Tsuite. Journal of the Mining and Metallurgical Institute of Japan. 1968;**965**:1394-1397. (In Japanese). DOI: 10.2473/shigentosozai1953.84.965_1394

[23] Ami N. On the fully mechanized mining of thin seams in Kushiro coal mine. Journal of the Mining and Metallurgical Institute of Japan. 1965;**926**:526-534. (In Japanese). DOI: 10.2473/shigentosozai1953.81.926_526

[24] Endo M. Taiheiyou Tankou ni Okeru Hakusou Saitan ni Tsuite. Colliery Engineering. 1966;**21**:66-72 (In Japanese)

[25] Taiheiyou Tankou Roudou Kumiai. Taiheiyou Tankou Roudou Kumiai Sanjuunen Shi. Kushiro: Taiheiyou Tankou Roudou Kumiai; 1976. p. 654 (In Japanese)

[26] Nakajima S. Experimental Research on Powered Support at Longwall Face: Changes of coal mining technique in Japan (1st Report). Shigen-to-Sozai. 2001;**117**:891-900. (In Japanese). DOI: 10.2473/shigentosozai.117.891

[27] Nakajima S. Observation and analysis of roof lowering wave observed at mechanized longwall face: Changes of coal mining technique in Japan (2nd Report). Shigen-to-Sozai. 2002;**118**:229-235. (In Japanese). DOI: 10.2473/shigentosozai.118.229

[28] Sasaki T, Takasaki M. Taiheiyou Tankou ni Okeru Kikaika to Shuuyaku Saitan ni Yoru Taryou Shuttan ni Tsuite. Colliery Engineering. 1965;**20**:194-201 (In Japanese)

[29] Sasaki T. Taiheiyou Tankou ni Okeru Chuukousou Kanzen Kikaika Saitan no Suishin. Toubu Tankou Gijutsu. 1966;**43**:106-114 (In Japanese)

[30] Yagi Y. On the Process of Developing a Self-advancing Face Support for the Coal of Medium Thickness in Taiheiyo Colliery. Journal of the Mining Institute of Hokkaido. 1966;**121**:69-77 (In Japanese)

[31] Takasaki M. Taiheiyou Kushiro Tankou ni Okeru Jisoushiho ni Tsuite. Tankou Shin Gijutsu. 1967;**16**:199-206 (In Japanese)

[32] Nakajima S. Taiheiyou Tankou Monogatari. Tokyo: Self-publishing; 2020. p. 412 (In Japanese)

[33] Tsuruoka Y. Chitei wo Kaeta Otokotachi. Tokyo: Mirai Bunkasha; 1999. p. 237 (In Japanese)

[34] Mitsui Miike Machinery. Sanjuunen no Ayumi. Tokyo: Mitsui Miike Seisakusho; 1989. p. 159 (In Japanese)

[35] Kishimoto Y. The development of overall high-speed workings by the SD system. Journal of the Mining and Metallurgical Institute of Japan. 1969;**974**:401-409. (In Japanese). DOI: 10.2473/shigentosozai1953.85.974_401

[36] Tamaru T. Development of Shield Type Self-Advancing Support. Journal of the Mining and Metallurgical Institute of Japan. 1972;**1013**:395-400. (In Japanese). DOI: 10.2473/shigentosozai1953.88.1013_395

[37] Murai I. Taiheiyou Kushiro Tankou ni Okeru SMK Jisouwaku ni Yoru Saitan. Colliery Engineering. 1970;**25**:194-198 (In Japanese)

[38] Kushiro Coal Mine. Shinbuka ni Tomonau Koudou Shiho Gijutsu no Hensen Chousa. Kushiro: Self-publishing; 2005. p. 116 (In Japanese)

[39] Taiheiyou Tankou Kanrishoku Kushiro Kurabu. Taiheiyou Tankou no Saikou Gijutsu. Kushiro: Self-publishing; 2005. p. 82 (In Japanese)

[40] Souritsu 50 Shuunen Kinen Jigyou Jikkou Iinkai, editor. 50 Nen no Ayumi. Tokyo: Mitsui Miike Mashinery; 2010. p. 233 (In Japanese)

[41] Bobson D, Takahashi K. Sausu Burai Tankou ni Okeru W-SD Shisutemu no 2 Men'me Saikutsujou deno Seika to Mondaiten. Journal of the Mining and Metallurgical Institute of Japan. 1977;**1076**:681-684. (In Japanese). DOI: 10.2473/shigentosozai1953.93.1076_681

[42] Shimanishi T, Ishikawa T. Taiheiyo Tankou no Heizan to Kushiro Kooru Ma'in no Hossoku. In: Shimazaki N, Nakazawa H, Shimanishi T, Ishikawa T, editors. Taiheiyou Tankou. Vol. 1. Hokkaido: Kushiroshi Kyouiku Iinkai; 2018. pp. 135-170 (In Japanese)

[43] Japan Organization for Metals and Energy Security. A report of the project for overseas dispatch training program for the advancement of the coal mining and safety technology in the coal producing countries [Internet]. 2014. Available from: https://coal.jogmec.go.jp/content/300274112.pdf [Accessed: December 1, 2022] (In Japanese)

[44] Japan Organization for Metals and Energy Security. A report of the project for training program in Japan for the advancement of the coal mining and safety technology in the coal producing countries [Internet]. 2014. Available from: https://coal.jogmec.go.jp/content/300274113.pdf [Accessed: December 11 2022] (In Japanese)

[45] New Energy and Industrial Technology Development Organization. Fact sheets 2007 [Internet]. 2008. Available from: https://www.nedo.go.jp/content/100098018.pdf [Accessed: December 11 2022](In Japanese)

[46] Kikkawa T. Nihon Denryokugyou Hatten no Dainamizumu. Nagoya: Nagoya University Press; 2004. p. 600 (In Japanese)

[47] J-POWER. Karyoku Hatsuden Jigyou ni Tsuite [Internet]. 2022. Available from: https://www.jpower.co.jp/bs/karyoku/kar00100.html [Accessed: November 26, 2022] (In Japanese)

[48] Japan Institute of Invention and Innovation. Sengo Nihon no Innovation 100 Sen [Internet]. 2016. Available from: http://koueki.jiii.or.jp/innovation100/innovation_detail.php?eid=00101&age=present-day [Accessed: November 26, 2022] (In Japanese)

[49] Anonymous. Integrated coal gasification combined cycle. The Thermal and Nuclear Power. 2020;**769**:601-610 (In Japanese)

[50] J-POWER. J-POWER group integrated report 2022 [Internet]. 2022. Available from: https://www.jpower.

co.jp/english/ir/pdf/2022.pdf [Accessed: November 26, 2022]

[51] New Energy and Industrial Technology Development Organization. Development of technologies for carbon recycling and next-generation thermal power generation [Internet]. 2022. Available from: https://www.nedo.go.jp/english/activities/activities_ZZJP_100115.html [Accessed: November 26, 2022]

[52] Murai R. Kaigaitan ga Nihon wo Sukuu. Tokyo: Kawade Shobo Shinsha; 2003. p. 259 (In Japanese)

[53] Ito T. Dynamism of J-POWER, national power company: A study of J-POWER's business model of large-scale, overseas-coal-fired thermal power station. Japan Business History Review. 2019;**53**(4):3-27. DOI: 10.5029/bhsj.53.4_3 (In Japanese)

Section 2

Statistical Tools Application

Chapter 3

Factor Analysis for Technology Management and Its Effectiveness in Indian Opencast Coal Mining

Anand Pd Sinha, Neha Choudhary, Rahul Rai,
Ashok Kumar Asthana and Praveen Chandra Jha

Abstract

Coal holds utmost significance as a natural energy source propelling a nation's industrial sector growth. Besides refining coal quality through adept mine technology management, contemporary mining grapples with multifaceted challenges encompassing human resettlement, land reclamation, forest preservation, pollution abatement, and efficient logistics. The coal mining sector serves as a tangible example where technology management assumes real-world importance. Despite adopting state-of-the-art methodologies, open-pit coal mining trails global standards. India's coal industry faces persistent struggles in accessing suitable domestic coal, relying on imports despite considerable technological strides. Beyond augmenting production capacity or product innovation, technology innovation concepts empower Indian enterprises to reshape their industries. Technology management research within mining remains in its infancy, necessitating a comprehensive grasp of its implications on internal operations and strategic alignment for global competitiveness and effective technology leadership. This study aims to dissect the integral facets imperative for proficient technology management within opencast coal mining domains.

Keywords: opencast mining, Indian coal sector, factor analysis, technology effectiveness, technology management

1. Introduction

The most crucial natural energy source for the expansion of any economy's industrial sector is said to be coal. With ever-increasing industrialization, India's need for electricity generation is expanding astronomically. The use of appropriate technology is required by the changing corporate environment of today to increase productivity. Almost all industrial sectors nowadays require efficient use of technology to maintain sustainability and competitiveness [1]. From a technological perspective, the Indian open-pit coal industry is at a transitional stage. Although the government now controls the bulk of coal blocks and mines, private corporations have joined the market and are giving public-sector organizations a difficult fight [2]. Technology has become increasingly important in the age of globalization and has sped up the pace of competition [3, 4]. Modern technology is essential for sustaining quality standards in a business environment. In the current Indian business environment, two technological components are crucial:

- Selection of appropriate technology
- Effective management and its utilization of proper technology

After pro-market reforms, technology has become the foundation of business sustainability. Technology management is a practise that involves categorizing, choosing, and implementing the technologies required to ensure an organization's continued existence and growth [5]. Despite significant investments in technology, the manufacturing sector still falls short of expectations in terms of technological performance [6]. Despite the adoption and usage of the latest technology, India's open cast coal mining industry is one of the key areas that lags behind in comparison to international standards. Modern technology is currently required by the open cast coal mining industry however, its installation and efficient administration are problematic.

From the perspective of technology management, the current study makes an effort to analyses the stated issue of low coal productivity [7]. The research makes an effort to examine many elements necessary for successful management of technology to increase the productivity of coal in CCL. It does this by using a structured questionnaire to collect primary data. The findings emphasize and carry out this effort. The study also tries to offer some guidelines for handling tools and using technology correctly and productively [8].

According to Khalil [9], appropriate technology is a suitable fit between the resources needed and the technology being used. Simply adopting new technology is a challenging endeavor because there are so many domestically and globally available alternatives. To comprehend good technology management and fully capitalize on it, the second factor has to be given greater attention [10, 11].

The notion of acceptable and inappropriate technology is shown in the image above. Any technology is appropriate at the time of development in relation to the environment for which it was created and in line with the primary purpose for which it was created. Because the environment and/or objective functions may have changed, it might or might not be acceptable at the same place at a different time [12]. Similar to this, it might or might not be acceptable at a different location at a different time, or at multiple times, depending on the surroundings and desired function. Therefore, technical appropriateness is not an inherent property of any technology, but rather arises from the context in which it will be employed as well as from the primary function (**Figure 1**).

Technology implementation and planning refers to the degree to which an organization has strategically planned the deployment of new technology(s) prior to its implementation, and the processes incorporated within this design, which influence the overall effectiveness of technology deployment and utilization. Internal planning and its implementation are a concern in the mining sector, which has an impact on coal output. Effective management throughout the implementation phase means assisting the project team, choosing the proper technology, and creating or giving the necessary training. This approach guarantees that new technology will enhance old procedures and that overall productivity will increase [14].

The manufacturing sector or industry of coal mining may be used as an illustration, where technology management can be seen as a practical problem. Open cast

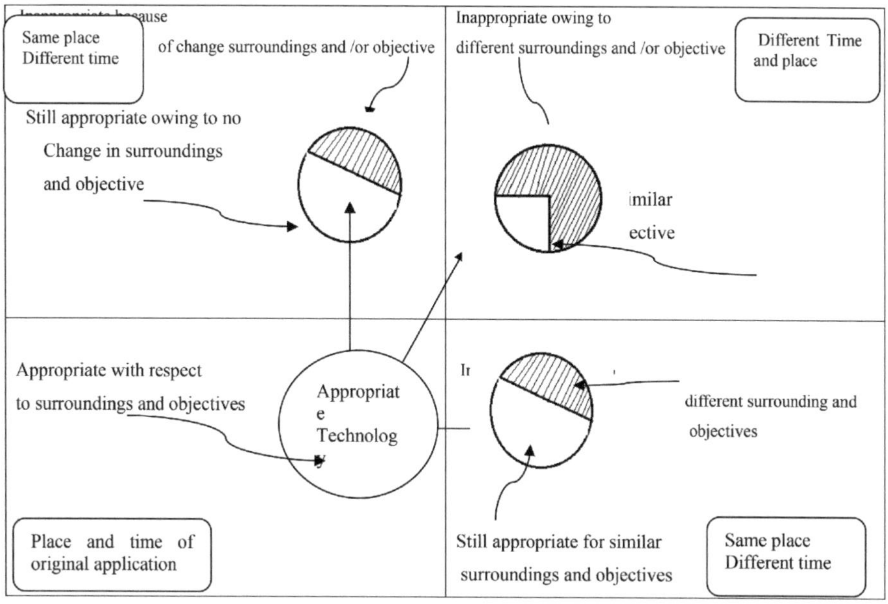

Figure 1.
Appropriate and inappropriate technologies. (Source: [13]).

coal mining employs cutting-edge technology, yet it still lags far behind in terms of worldwide standards. Despite several technical advancements, the Indian coal industry still struggles to obtain coal that is acceptable for home use rather than relying on imported coal [15]. Without suitable organizational adjustments as well as improvements in human capabilities, technology cannot be successful. Researchers believe that government-controlled coal mining enterprises lack appropriate technological management and are unable to demonstrate returns on investment (**Figure 2**).

Although technology cannot alter things on its own, it can when it is accompanied by sensible actions and well-developed human skills. There is no doubting that selecting the improper technology has negative effects on the organization's overall health, but the key is to manage technology well as well [17]. This research makes an

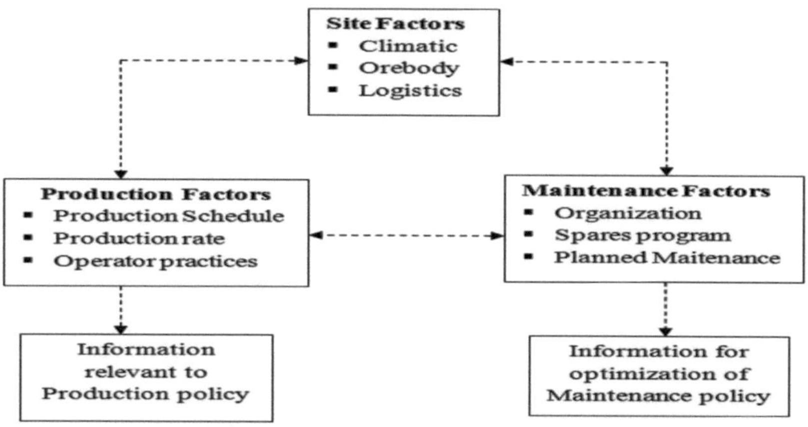

Figure 2.
Factors affecting maintenance strategies [16].

effort to identify issues related to the adoption, deployment, and effective use of the installed technology. This study's background is based on managerial concerns with regard to the efficient management of technology. One of the major public sector coal companies, Coal India Limited, together with its seven open-pit subsidiary mines and its designing division, are the subjects of the data gathering [18, 19].

This study primarily focuses on identifying the causes of the low coal productivity in CIL's seven mines, namely WCL, ECL, MCL, CCL, NCL, SECL, BCCL, and CMPDIL. This study aims to investigate the many elements needed for efficient technology management in opencast coal mining sectors. After pre-testing, reliability, and validity, the data to be collected is placed into statistical analysis, and variables affecting technology management are eventually found by utilizing Factor Analysis in Mathematical Software, i.e., Statistical Package for Social Science (SPSS) platform.

1.1 Problem identification

The most significant energy source for the production of power and for sectors like steel, cement, fertilizers, and chemicals is recognized to be coal. Therefore, the Indian coal sector requires greater investment, and private companies' active participation is also required to improve output level in order to meet the need for coal. There is no assurance that coal will be supplied at a specific quality (size, ash content, calorific value, etc.), and there are no consequences for breaking the rules either. Despite possessing a large coal deposit, imported equipment and technology, as well as a large market for coal, real coal output falls short of the desired level (**Figure 3**).

Despite having such a sizable natural coal supply, output falls short of expectations. In the open cast coal mining sector, coal is mined using HEMM (primarily drill machine, shovel, and rear discharge dumper) in combination. It is comparable to a track and relay race in which no runner competes on their own and every participant contributes to the success of the race in some way. Despite the fact that they are all on the same team, the second runner's performance in the race and his ability to successfully transfer the baton to the next team member depend on how well the first runner (the drilling machine) delivers the baton to him (the shovel). These hand-offs affect not only the next leg of the race (Rear Discharge Dumper), but the success of the entire team. And for smooth hands-off, efficiency and productivity of individual HEMM is the key. Thus, HEMM need to work in consonance with each other to optimize the productivity of the system.

The rationale can be explained mathematically as illustrated below:

Let us assume,

$$m \quad n \tag{1}$$

$$A \rightarrow B \rightarrow \tag{2}$$

$$At = A_0\, e^{-mt} \tag{3}$$

$$Bt = A_0{}^{*}\ [m/(n-m)]^{*}\ e^{mt}-e^{nt} \tag{4}$$

$$Ct = A_0\ [1 + \{1/(m-n)\}^{*}\ (ne^{-mt}-me^{-nt})] \tag{5}$$

Here, A represents Drill, B represents Shovel and C represents rear discharge dumper. A_0 represents the initial function being performed by the drill. At represents number of pockets/holes to be drilled by the drill machine at time t.

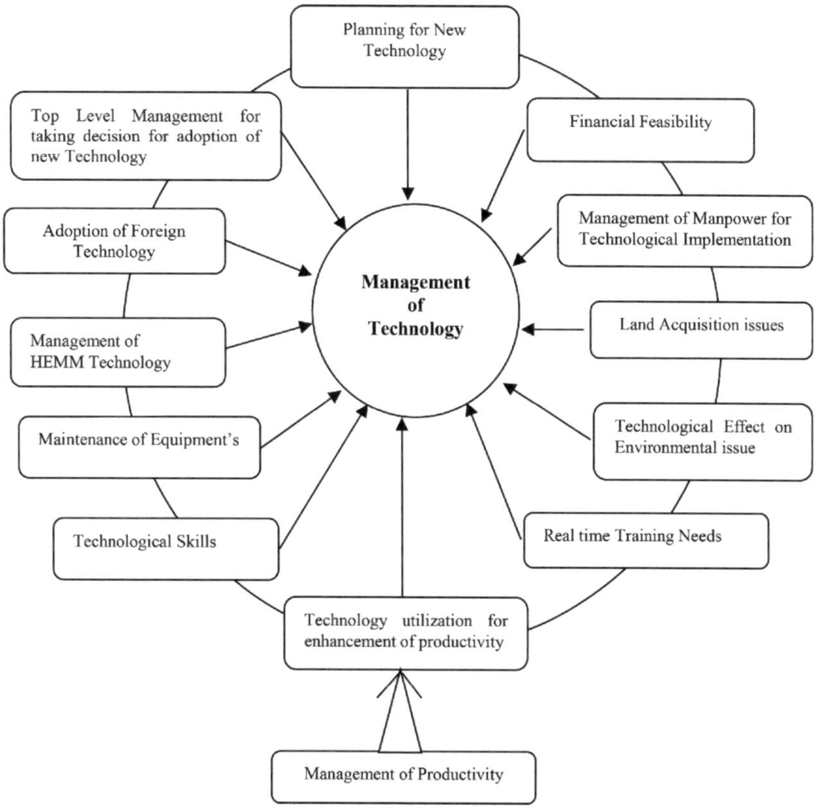

Figure 3.
Flower model with extracting variables.

Bt represents net material available with the shovel at time t and Ct represents net material loaded on the rear discharge dumper at time t. Clearly, the net material loaded on the dumper at any time instant t depends on the kinetic factors of the drill (m), the kinetic factor of the shovel (n) and the initial material/ area available with drill (A0) The above formula is an analogous and applied version of the concept of series chemical reactions used in chemical kinetics. For the present work, three critical elements quintessential for productivity enhancement in open cast coal mining w.r.t. technology have been taken into consideration [20]. These are Drills (Element A), Shovels (Element B) and Rear Discharge Dumper (Element C) [21]. These three elements need to work in series in consonance with each other to optimize the productivity [22]. Being a series operation, individual efficiency of the predecessor constituent affects the next constituent. The net material loaded on the rear discharge dumper at any time instant t depends on the kinetic factor of the shovel (n), the kinetic factor of the drill (m) and the initial material/area available with drill (A0). Clearly, individual efficiency of the predecessor element affects the next constituent and hence the success of the system.

In order to provide appropriate corrective actions through perspective policies and suggestions, the study work raises a variety of challenges and concerns linked to managing technology and its efficacy in the coal mining sectors, namely Coal India

Limited and its subsidiaries. India ranks third in the world for coal production. It has increased coal output from 70 Mt. at the time of nationalization in the early 1970s to 355 Mt. (provisional - excludes Meghalaya) in 2009–2010 through a consistent investment programme and a stronger emphasis on the deployment of new technology. Over 81% of India's coal is produced in open-pit mines, which contribute to the country's overall production. Despite this, the industry remains unappreciated. The most significant source of energy for the production of electricity is acknowledged to be coal [23]. In addition, many small- and medium-scale businesses depend on coal for their operations and energy needs, including those in the steel, cement, fertilizer, chemical, and paper sectors. Many mining and industrial organizations look for ways to save expenses and eliminate overhead by producing more with fewer resources. The manufacturing industry has made tremendous strides in integrating new technology into its processes, including advanced manufacturing and Industry 4.0 [24, 25].

The mining sector, in comparison, is still lagging behind when it comes to integrating modern technologies into its operations. Despite such a persuasive position and demand, there are serious concerns about the output of coal from CIL. Following are some categories of plausible causes for such a poor performance [26]:

- Despite having a current fleet of equipment, production was lower.
- Inappropriate management and the use of cutting-edge technology might result in a problematic scenario. Overdependence on foreign technologies might emerge from this. It is being made worse by supplier project delays and other associated issues.
- CIL must intensify its methods for raising the quality, quantity, and cost-effectiveness needs of its clients in order to thrive in the current industrial environment.
- The indicated company's use of antiquated mining technology with reduced capacity;
- Cost and time overruns caused by a lack of structural, tactical, and strategic difficulties.
- The output has also been significantly impacted by the following factors: the use of smaller fleets of trucks, conventional equipment, and shovel equipment rather than draglines and bigger sized fleets; the lack of skills necessary for adequate planning, supervision, and management; and the problem itself.
- Ineffective management and a lack of operational will have also exacerbated the issue. This problem in regard to the technology management component has also been made worse by improper and insufficient examination of the environmental, social, economic, and community implications, as well as by actual and meaningful involvement with stakeholders.
- Some of the major issues with open cast mines, such as those at the Piparwar and Ashoka mines operated by CCL, are also connected to the halting of coal transportation as a result of siding and other illegal actions.

- A lack of law and order serves as a stimulus for lower coal output.
- The difficulty in acquiring environmental and forestry clearances is delaying a lot of mining operations, and the cost and time overruns caused by a delayed clearance have made the issue of decreased production even worse.
- Surface mining necessitates the disturbance of a comparatively bigger area of land. The local biodiversity and ecological balance are disrupted as a result of numerous environmental problems like soil erosion, water pollution, and others, and occasionally local populations and environmental protection organizations also present operational challenges, which ultimately have a negative impact on production.
- Other factors contributing to the actual production of coal falling behind include inadequate drilling capacity, a backlog in overburden removal, an imbalance between excavation and transportation capacities, poor availability and under-utilization of HEMM, etc.

1.2 Need for the study

After conducting exploratory research in Coal India Ltd. following points have emerged as issues, which itself highlights the need and urge of the present study:

- Targeted production is higher as compared to actual production repeatedly has urges the need to investigate the issues.
- Despite of having the modern equipment's and technologies their management is not proper and the capacities of installed technology are underutilized.
- Preliminary data study revealed that the installed technology has little significant effect on the production and management of installed technology.
- Forgoing and nothing to bother attitude of management as well as worker.
- The urging importance of coal industry as the most important source of energy in current scenario.
- Personal interest and attraction towards the coal mining sector has also added to the need of study.
- Lack of adequacy of training programme and industrial relation policies at the selected sites of CIL.

1.3 Objective of the study

Objective of this proposed study may be highlighted and identified in the following directions:

1. To understand the importance of Managing Technology and its effectiveness in Open Cast Coal Mining Industry.

2. To find out the problems related to adoption and implementation of technology in Mining Industry that effect production of coal.

3. To determine the numerous factor necessary for efficient technology management in order to increase coal productivity.

2. Literature support: importance of technology and its effective management

Technology management is a field that combines the use of science, engineering, and management skills to fulfill an organization's technology needs. The life cycle of various technologies is handled through technology management in order to fulfill organizational goals. Understanding the value and suitability of a technology for an organization is, thus, the primary responsibility of technology management [27]. An opportunity in the market will soon end, thus a sustainable organization will try to spot it as soon as possible and take use of it for the project that must be completed quickly [28].

The effectiveness of technology also depends on non-technical factors. Non-technical variables also affect how well technology works. The same tool or machine can be operated incorrectly or correctly. The mere possession of technology is useless unless it can be handled proficiently and skilfully, which necessitates knowledge of and aptitude for industry-specific technology management [29]. Many advantages may come from managing technology, but much will rely on how those who are engaged use it. Therefore, organizations must have both management and technical experts. Technical skill is defined as the capacity of an individual to commence and finish a specific work or job using the tools and procedures proficiently [30].

Although many people are technically adept, their interpersonal skills are often lacking. Managers must possess the best possible blend of all necessary managerial skills, be able to delegate, and handle complex and contentious situations diplomatically because they are the operational and symbolic head of an organization and are responsible for tasks and results [31]. The efficient accomplishment of a well-defined set of objectives is another crucial component, which is referred to as team effectiveness. By becoming sensitive to the rapidly changing internal and external environments, successful organizations continually strive to increase the performance of their teams. This project calls for a variety of abilities and capabilities that must be supplemented among the team members. Some of the literature supports are mentioned below (**Table 1**).

2.1 Management of technology - leading variables

This study is expected to be conducted on installed technologies managed by Coal India Ltd. and its subsidiaries. Extensive literature, annual and other statistical report published by Coal India Ltd. was studied and analyzed properly to know about the leading variables involved in management of technology. It was identified that actual production of coal was below the targeted. CIL is known for its market capitalization and technological up gradations efforts but despite of this production for the considered time period was less. This has given a platform to develop hypothesis of research work in context to effective and efficient management of technology. Important

Contents	Author	Year	Remark
Knowledge of maintenance techniques is required to increase safety and output capacity in mining, and this knowledge should be based on the fields of interacting variables to maintenance.	Watson	[16]	Maintenance of overall Equipment's
Maintenance staff must get training in their specific working environment, and this training should be organized such that each employee is informed of the most recent maintenance difficulties and approaches. Safety training is crucial as part of the training requirements.	Edwin B. Flippo	[32]	Real time Training for Technical up-gradation
The operational element of the organization deals with its daily operations, whilst the planning factors concentrate on its long-term problems.	Ghatak	[33]	Operational Performance
It has been suggested that the procedures included in this design have an impact on how well technology is deployed and used overall.	Bancroft, Haddad,	[34]	utilization of technology
A process of surface coal mine planning involves selection of coal property, making decisions regarding appropriate mining method, selection of types and number of equipment's; and producing mine designs to make an optimum use of equipment's and manpower.	Steele	[17]	Planning for New Technology
The discovery and assessment of technologies, the development of new or enhanced goods and processes, the integration of technology with other business processes, and the management of change necessitated by the adoption of technologies all need complicated judgments from effective management.	Allan C. Wexler	[2]	Technological Performance
There are limitations and a number of factors that contribute to this, particularly problems with land acquisition, forest and law and order issues, as well as evacuation issues, have hampered coal production to the point where even captive coal blocks have fallen short of production goals.	Saxena	[35]	Technological barrier due to Land Acquisition
If not adequately planned and managed, mining activities cause significant environmental and ecological harm. Technology will inevitably be used to increase yield.	Boskin, M. J.	[36]	Real time Technological Advancement
The choice of technology has a significant impact on industrial growth and productivity. Technology utilization is always constrained by a goal.	Khalil	[37]	Adoption of Foreign Technology /Selection of technology is prime importance
The transformation that technology brings about must be supported by suitable organizational reforms, changes in human capabilities, and changes in training and education.	Christina Beach	[38]	Technological Skills
The management of HEMM (Heavy Earth Moving Machineries), such as the shovel-dumper combination model, aims to integrate two fundamental pieces of equipment in order to increase production and efficiency.	Ghatak. S	[39]	Management Of HEMM

Table 1.
Literature support.

inferences that can be drawn after the analysis of report present a very interesting fact that management of installed technology is affected by some factor. Technology installation is not only sufficient for organization but its effective management does matters. Pertaining to the research some variables which were important from point of view of management of technology were identified and from them factors were extracted. These variables associated with all the stages of maintenance of technology ranging from its planning stage to implementation stage. This study identified various leading variables are (**Table 2**) [40].

After a thorough review of the literature, comprehensive interviews with shop floor managers, candid discussions with top level decision-makers within the industry, as well as initial observations and primary research, the total of 24 leading variables were found (**Table 3**).

Sl. No.	Effecting Mgmt. of Technology
1.	Planning for new Technology
2.	Selection of Indigenous Technology
3.	Selection of Foreign Technology
4.	Technological Skills
5.	Financial Feasibility
6.	Cost and Benefit Analysis
7.	Real time Technological Advancement
8.	Managing HEMM Technology
9.	Supply chain issues
10.	Waste reduction by applying new technology
11.	Real time transfer of technological change8
12.	Socio-Economic issue on new Technology
13.	Maintenance of overall Equipment's
14.	Continuous Monitoring of Quality
15.	Proper Utilization of Machines
16.	Real time Training for Technical up-gradation
17.	Safety needs for continuous technology
18.	Top Level of Mgmt. for adoption of new technology
19.	Middle Level of Mgmt. for adoption of new technology
20.	Technological barrier due to Land Acquisition
21.	Technological Effect on Environment Issues
22.	Proper Management of Manpower
23.	Market Feasibility
24.	Policy Implications

Table 2.
List of leading variables for effective Management of Technology.

Sl. No.	Effecting mgmt. of Technology	Lit. support lead by Authors'/ Practitioners'
1	Planning for Technology	Steele LW, [17]
2	Selection of Foreign Technology	Khalil, [41]
3	Selection of Indigenous Technology	Khalil, [37]
4	Technological Skills	Christina Beach, [38]
5	Financial Feasibility	Betz, Fredrick [42]
6	Cost and Benefit Analysis	Boskin and Lau, [36]
7	Real time Technological Advancement	Mehta, [43]
8	Managing HEMM Technology	Ghatak, [39]
9	Supply chain issues	Monika Maria, [44]
10	Waste reduction by applying new technology	Allan C. Wexler, [2]
11	Real time transfer of technological change	Moustafa, M. E, [45]
12	Socio-Economic issue on new Technology	Stewart, [46, 47]
13	Maintenance of overall Equipment's	Watson, [16]
14	Continuous Monitoring of Quality	Sevim & Lei, [48]
15	Proper Utilization of Machines	Bancroft, [34]
16	Real time Training for Technical up-gradation	Edwin B. Flippo, [32]
17	Safety needs for continuous technology	J. Ritson, [49]
18	Top Level of Mgmt. for adoption of new technology	Tarek Khalil, [37]; Pal et al., [50]
19	Middle Level of Mgmt. for adoption of new technology	Tarek Khalil, [37]; Koontz, [51]
20	Policy Implication	Ghatak, [33]
21	Technological barrier due to Land Acquisition	Saxena, [35]
22	Technological Performance	Singh Gurdeep, [52]
23	Proper Management Of Manpower	Chhipa et al., [53]
24	Market performance	Berman E. M, [54]

Table 3.
Literature support.

3. Coal India limited: a overview

In the industrial economy of the nation, coal has gone a long way to become one of the main sources of energy. The government took control of non-coking coal mines on January 31, 1973, and the Coal Mines Authority Limited was established with four operational divisions, including the Central Division of CMAL, which included NCDC. Further coalmine restructuring led to the establishment of CIL as the controlling company in 1975. Coal and coal products are produced and supplied by CIL and its subsidiaries to key industries such steel, power, cement, fertilizers, defense, and railways. CIL has eight subsidiary companies; details can be viewed as (**Table 4**) [55]:

Coal India Limited has acquired the status of the third largest coal producing company of the world, having its noble start in the year 1975 as a holding company,

Company	Headquarters	Year of corporation
Eastern Coalfields Limited (ECL)	Sanctoria (WB)	1975
Bharat Coking Coal Limited (BCCL)	Dhanbad (Jharkhand)	1973
Central Coalfields Limited (CCL)	Ranchi (Jharkhand)	1975
Northern Coalfields Limited (NCL)	Singrauli (MP)	1986
Western Coalfields Limited (WCL)	Nagpur (Maharashtra)	1975
South Eastern Coalfields Limited (SECL)	Bilaspur (MP)	1986
Mahanadi Coalfields Limited (MCL)	Sambhalpur (Orissa)	1992
Central Mine Planning and Design Institute Limited	Ranchi (Jharkhand)	1975
NorthEastern Coalfields Limited (NECL)	Meghalaya	1975

Table 4.
CIL with eight subsidiaries.

under ministry of coal, the company is now a maharatna company. The company is responsible for the production of 90% of the coal requirements of India. Captive Mines of TISCO, IISCO and DVC are also related to it. Coal India currently operates 510 mines and 15 washeries spread over nine states toproduce and beneficiate coal for meeting the demand of the consumers all over the country. (*Source: Coalindia.nic.in*) (**Table 5**).

3.1 Evolution of technology - coal mining industry

There are a few fundamental aspects of technology. According to Mashelkar [56], it is largely an ideational process that uses ideas to change both the material and non-material worlds. Technology is behavioral because it calls for the use of skills in both tool invention and tool usage. It is organizational and institutional since it is culturally ingrained and in opposition with the stifling institutional values. Since it is possible to combine, recombine, and change already existent technology once the process of technology has begun, it is cumulative and combinational [57]. It is a collaborative approach that incorporates social interactions and feedback loops. Technology accelerates both the problem-solving and the process. In contrast to an evolutionary process, it always enables individuals to both accomplish new things and do old ones better. The 1950s and beyond saw the development of contemporary management theories as well as organized efforts in the field of technology management (the era was distinguished by an abundance of resources for R&D).

In the 1970s, management of innovation began to operate, and the business world as a whole became interested in understanding innovation and how it should be used. However, development slowed down in the twentieth century as a consequence of the effects of global competition and the American economic crisis. Mechanical rock cutting equipment was first introduced to the opencast mining sector in the early 1980s. Lignite, coal, limestone, and gypsum were the first materials used [58]. Drilling, blasting, loading, crushing, and transportation processes are always included in traditional opencast mining operations. The effectiveness of drilling and blasting has a significant impact on the efficiency of operations. All around the nation, opencast mining searched for machinery that may help to solve these issues [59].

Sl NO	Company	2014–2015 (Mt)		2015–2016(Mt)		2016–2017(Mt)		2018–2019(Mt)		2019–2020(Mt)		2020–2021(Mt)	
		Target	Actual	Target	Actual	Target	Actual	Target	Actual	Target	Actual	Target	Actual
1	ECL	22.57	22.20	23.18	15.74	20.34	19.74	21.75	21.83	24.20	23.20	25.19	24.05
2	BCCL	19.59	19.30	20.62	20.75	21.50	21.38	23.45	23.61	24.75	25.31	25.19	25.30
3	CCL	42.00	42.32	44.90	41.68	52.6	45.61	62.60	47.52	75.60	48.00	75.50	48.05
4	NCL	52.00	52.16	58.00	59.62	61.25	63.65	66.50	67.67	72.00	66.25	75.55	68.25
5	WCL	32.10	33.30	32.39	33.53	32.75	34.59	34.85	36.12	36.35	34.95	36.95	35.65
6	SECL	71.00	72.30	74.04	77.05	78.00	83.58	88.50	90.18	93.50	95.90	94.50	95.37
7	MCL	77.59	78.03	85.60	85.89	96.11	94.19	107.20	101.88	114.46	98.11	115.35	101.95

(Source: Project and Planning Department- CMPDIL, Ranchi).

Table 5.
Production of coal eight subsidiaries under CIL.

The advent of continuous surface miners in the early 1990s provided a solution to these issues in Europe and internationally, marking the beginning of environmentally friendly mining practises [58]. It is a practical substitute for rock breaking that does away with drilling, blasting, loading, and crushing processes. It could resolve grievances brought up by these actions. In 1993, India's first surface miner was launched as a result. In India, surface miners have been effectively used in coal and limestone mines. It has now been shown that this technology is groundbreaking for our time. The first time a surface miner was used was in the Lakhanpur opencast project, operated by Mahanadi Coalfields Limited, a division of Coal India Limited. At 2006, Central coal filed ltd. also began using surface miner at its Ashoka opencast coal mines.

The researcher had the chance to evaluate a surface miner's performance. Due to a village's close vicinity, the notion of employing or using a surface miner at mining activities was born. Over a five-year period, 700,000 tonnes of coal were blocked. The Lakhanpur opencast project's successful use of a surface miner led to better quality through selective mining and environmentally sustainable coal production. Additionally, it encouraged the commercial and public mining industries in India to employ this adaptable machinery more frequently in order to satisfy their need for coal.

Through its own conveyer boom, this machine cuts and loads coal. This machinery removes first and even secondary crushing in mineral and rock deposits as an alternative to traditional drilling and blasting operations. In situations when drilling and blasting are not feasible, surface mines are a specialized mining technique that is frequently utilized. This machine does not require drill and blast or subsequent crushing as the cutting drumps break and size rock. These machines can discharge onto conveyor belts or directly load truck or work in windrowing mode in which machines cut the material and leave the material on the floor and cut face as it to be loaded by small size front-end loader on small size dumpers for transport from mines to destination point. Generally machine requires a large area of exposed coal for efficient operation. Size of mined coal is such that further crushing is generally not required. The thin layer of coal is taken at a time, the machine is capable of cutting and loading medium hard dirt bands separately. This machine can be equipped with a sensing system to detect and identify different materials by measuring infrared radiation reflected by mineral deposits. These readings allow adjustment of cutting depth for selective mining. With use of this machine, coal washing for removing obvious dirt from R.O.M (Run of Mine) coal can be eliminated. A washery will be much costlier, both in capital and running costs, then a set of surface miners. In 1993, Piparwar project, an Indo-Australian venture has been carved out to develop a new coal mine with beneficiation plant for non-coking coal to meet the demand of power coal of consistent quality. The project is designed to achieve a very high level of productivity through introduction of Mobile Inpit Crushing and Conveying Technology in subsidiary of Coal India Ltd., i.e., Central Coal Fields Ltd. This project was started on a bilateral agreement basis. Government of India requested the Australian Government's involvement to develop the Piparwar opencast project.

3.2 Technology: innovation and its effective management

The successful use and integration of technology within an organization is crucial. A wide range of activities, data, and skills must be coordinated for project conception and execution to be successful. Due to the fact that commercial possibilities are time-limited, an organization must move swiftly in order to take use of cutting-edge technology effectively for projects that must be completed rapidly [60]. These obstacles in the corporate environment have increased the demand for efficient technology management and control. As a result, achieving any objective, whether at the corporate or personal level, demands a methodical and carefully thought-out decision-making process. Clear objectives must be defined in order for management to function successfully or efficiently [61].

3.2.1 Maintenance of overall equipment's after adoption of new technology

Technology management is essential to maintenance work. The total of all technical, administrative, and managerial actions taken to retain or restore an item to a condition where it can carry out the required function constitutes maintenance during the course of an item's life cycle. Any form of machinery that is used requires regular maintenance and repairs. Because of the environment in which mining production systems work, safety assurance is a vital factor that must be carefully considered when dealing with operational company entities [62]. To boost safety and production capacity in mining, understanding of maintenance procedures is necessary. This knowledge should be based on the fields of interacting variables to maintenance. Some of the most important, interconnected factors that have an impact on a mine production system's reliability Watson [16].

It is evident that equipment dependability will be enhanced quickly, operating costs will be reduced, and profit maximization will be the end outcomes if these interrelated aspects are controlled appropriately [16]. In addition to exercising the necessary management and technical control of maintenance programmes, procedures and strategy are often generated from maintenance management for all maintenance-related operations. The same way that business goals are communicated to other business organizations, it is typically vital for manufacturing or production businesses to establish, develop, and communicate the maintenance strategy. Regarding maintenance practices and procedures, [63] suggested that the maintenance management process has two parts: the first is effectiveness, which primarily deals with identifying the most significant problems and potential solutions, and the second is efficiency, which deals with identification of the suitable procedures. While using a participative method, management aspirations and expectations should be kept in mind. Every organization that operates in a setting of intense competition strives to succeed by increasing its efficacy. Non-technical variables also affect how well technology works. The same tool or machine can be operated incorrectly or correctly. Technology is useless unless it can be employed proficiently and skillfully, which calls for knowledge of and aptitude in technological administration by industry [64]. However, a committed and knowledgeable team of human resources is what drives performance in an organization. On the other side, productivity is the achievement of goals via the use of resources like money, labor, equipment, infrastructure, etc. It speaks to the interaction of inputs and outputs or the effectiveness with which organizational goals are accomplished. The efficient accomplishment of a well-defined set of objectives is another crucial component, which is referred to as team effectiveness. By becoming sensitive to the rapidly changing internal and external environments, successful organizations continually strive to increase the performance of their teams. This project calls for a variety of abilities and capabilities that must be supplemented among the team members.

4. Research methodology

Standard research methodology was adopted in context to the present work. Data sources and collection methods were carefully chosen and the self-administered primary data collection tool (questionnaire) was pre validated before data collection. Observation, Interaction and interview of the respondents were another tool used in this regard. Designing questionnaire was very comprehensive; it was carefully planned and designed after identifying different variable considered important pertaining to the present study. The core items and variables in the first stage were identified after conducting extensive literature survey (Published work, Journals, Company books and Annual reports etc.). The identified factors were supplemented by another set of variables discovered and marked on the basis of personal interactions and interviews. Senior management representative, employees and expert were then contacted to verify the appropriateness of variables identified and their flow. Personal interview helped a lot in understanding views and perception of the respondents related to efficient and optimal management of the installed technology. Secondary data from syndicate source has also been considered for collecting information to specific queries.

4.1 Hypotheses testified

The technology management system used at open-pit mines. The study's findings suggest that there is an organizational framework for technology management. The production department reports to the Director (Operation), who is in charge of planning, organizing, staffing, implementing, and controlling the production process. The General Manager (Operation) of the various mines assists the Director (Operation) in developing production plans and strategies. The Deputy General Managers of various sub divisions provide support to the General Manager (Operations). Regular updates are made to the production plan to account for evolving circumstances. The revision of the production schedule demonstrates the Director (Operations)' continued confidence in long-term technological planning to guide the organization along the intended course.

H_1 = Technology is not only a panacea for increasing production rather, efficient management of technology is crucial for increasing productivity.

4.2 Factor analysis

Output of Factor Analysis is obtained by requesting Principal Components Analysis and specifying a rotation using varimax. Eigen values associated with each linear component (factor) before extraction, after extraction and after rotation. Before extraction 24 linear components were identified within the data set. The Eigen values associated with each factor represents variance explained by that linear component and also displays the Eigen value in terms of percentage of variance explained. After factor analysis total 24 variables were reduced to 12 variables and from that 3 factors that were statistically significant were identified. In this case those variable, factor loading is more than percentage of communality only that value will be extracted and rest of the variable will be dropped. After calculating the factor analysis, the variables value is VAR6, VAR14, VAR 15, VAR13, VAR21 have high loading of .969, .900, .955, .916, .933, respectively on factor 1. This suggests that *factor 1* is a combination of above 5 variables.

Component	Initial Eigen values			Extraction Sums of Squared Loadings			Rotation Sums of Squared Loadings		
	Total	% of Variance	Cumulative %	Total	% of Variance	Cumulative %	Total	% of Variance	Cumulative %
1	6.762	28.174	28.174	6.762	28.174	28.174	6.590	27.458	27.458
2	6.277	26.156	54.330	6.277	26.156	54.330	6.196	25.815	53.272
3	5.852	24.384	78.713	5.852	24.384	78.713	6.106	25.441	78.713
4	.851	3.547	82.260						

Table 6.
Total variance explained.

At this point, the researcher's task is to find a suitable phrase which captures the essence of the original variables which continue to from the underlying concept or 'factor'. In this case, factor 1 could be named as *'Mine Planning & Design'* as shown in table no. 8. Similarly, for *factor 2,* variables VAR10, VAR11, VAR23 have a high loading of .883, .889, and .986, respectively, this indicates that factor 2 is a combination of the above 3 variables. In this case, factor 2 could be named as *'Evolution of Technology'*. For factor 3, variables V1, V4, V12, V22 have a high loading of .957, .884,.966, .935, respectively, this indicates that factor 3 is a combination of the above 4 variables. In this case, factor 3 could be named ass *'Effective Management of Technology'*. Out of 24 variables only 12 variables have extracted because of Rotation Sums of Squared Loadings percentage of cumulative value is .78 and rest of variables is low loading value, i.e. it has dropped for further analysis and only 12 variables will be applicable which is statistically significant. The analysis data has given below (**Tables 6–9**):

Codes	Variables	Factor to which a variable is merged	Factor Loading	Communality
VAR01	Top Level Management	Factor – 3	.957	.926
VAR02	Middle Level Management	Factor – 3	.814	.667
VAR03	Adoption of Indigenous Technology	Factor – 3	.748	.587
VAR04	Adoption of Foreign Technology	Factor – 3	.884	.783
VAR05	Market Feasibility	Factor – 1	.712	.632
VAR06	Financial Feasibility	Factor – 1	.969	.955
VAR07	Cost and Benefit Analysis (Economic feasibility)	Factor – 1	.814	.712
VAR08	Real Time Technological Advancement	Factor – 3	.829	.717
VAR09	Continuous Monitoring of Quality	Factor – 2	.814	.731
VAR10	Technology utilization for enhancement of productivity	Factor – 2	.883	.785
VAR11	Real time Training Needs	Factor – 2	.889	.804
VAR12	Management of HEMM Technology	Factor – 3	.966	.941
VAR13	Management of Manpower for Technological Implementation	Factor – 1	.916	.857
VAR14	Land Acquisition issues	Factor – 1	.900	.814
VAR15	Technological Effect on Environmental issue	Factor – 1	.955	.939
VAR16	Supply Chain Issue & Spare Parts management	Factor – 3	.792	.698
VAR17	Minimizing Wastage by applying New Technology	Factor – 2	.857	.744
VAR18	Socio-Economic Issue on New Technology	Factor – 1	.846	.726
VAR19	Real Time Transfer of Technological Change	Factor – 2	.849	.723
VAR20	Policy Implication	Factor – 1	.929	.864
VAR21	Planning for New Technology	Factor – 1	.933	.875

Codes	Variables	Factor to which a variable is merged	Factor Loading	Communality
VAR22	Maintenance of Equipment's	Factor – 3	.935	.889
VAR23	Technological Skills	Factor – 2	.986	.975
VAR24	Safety Need for Technology	Factor – 2	.736	.549

Table 7.
Summary results of factor analysis.

Factors	Factor-1 Mine planning and design	Factor-2 Evolution of Technology	Factor-3 Effective Management of Technology
Variables	Financial Feasibility (V6)	Technology utilization for enhancement of productivity (V10)	Top Level Management for taking decision for adoption of new Technology (V1)
	Land Acquisition issues (V14)		Adoption of Foreign Technology (V4)
	Technological Effect on Environmental issue (V15)	Real time Training Needs (V11)	Management of HEMM Technology **(V12)**
	Management of Manpower for Technological Implementation **(V13)**	Technological Skills (V23)	Maintenance of Equipment's **(V22)**
	Planning for New Technology **(V21)**		

Table 8.
Identified leading variables under factorization for Management of Technology.

Chi-Square Tests			
	Value	**DF**	**Sig. (2-sided)**
Pearson Chi-Square	17.443	4	.002
Likelihood Ratio	18.627	4	.001
Linear-by-Linear Association		8.886	1.003
No. Of valid Cases	380		
a. 0 cells (0%) have expected count less than 5. The minimum expected count is 11.29.			

Table 9.
Test of hypothesis.

The comments provided by the chosen respondents from open-pit coal mines showed that effective management of technology is crucial to boosting coal extraction productivity. Chi-square analysis yields a value of 17.443 at 4 degrees of freedom and a P-value of. 002, which is higher than the tabulated value at. 05 in 95% confidence level, indicating that the result is significant. Therefore, the null hypothesis is rejected, and the findings indicate that efficient management of technology may be essential to boosting output. Technology can only increase an organization's efficiency; but, if it is not managed properly, the entire goal is defeated. If technology is not managed well, it will have an impact on output.

4.3 Factorization and flower model

Significant variables are identified statistically, and the corresponding factors extracted by the researcher are important pillar of the work and are highlighted below:

- Mine planning and design is first step having an indispensable effect on productivity of mine and selection of technology. In context to the stated point researcher has identified Financial Feasibility, Land Acquisition issues, Technological Effect on Environmental issue, Management of Manpower for Technological Implementation, Planning for New Technology as variables. All stated variables are important to design the layout of mine. Layout and sequencing are having a significant effect on productivity.

- Evolution of technology is second most important factor identified. Many technology options are present in market in context to enhancing productivity in coal mine but the problem lies in its effective selection and implementations. Researcher has pointed out the following variables important in this regard, Real Time Training Needs, Technology utilization for enhancement of productivity, Technological Skills. All these variables mentioned are quite important in planning and evolution of technology as a mismatch between selection and skills will result in leakages and casts a negative effect on utilization of technology and motivation of employees.

- Effective management of technology is another important aspect to be taken care of, in absence of proper management and selected technology. Capacity of installed technology cannot be utilized to the fullest and frequent breakdown and ineffective utilization may be evident in the system. Researcher has identified important variables in relation to management of technology which are technology utilization for enhancement of productivity, Real Time Training Needs.

Flower model is an intuitive approach to enhance productivity and efficiency in an integrative manner. Variables identified are complimentary to each other and are not discrete rather they are interdependent and guided towards the epicenter of management of technology. Out of 12 variables, which were found significant subsequently, were extracted in three factors namely; Mine Planning and Design, Evolution of Technology and Effective Management of Technology. These factors play a significant role in Selection, implementation and management of technology in an integrative manner and finally have a positive effect on enhancing productivity in prime sectors like coal mining.

All variables identified by the researcher represents petals of flower as the petals are joined together to the base of flower in a similar fashion these variables are coordinated, combined, interrelated and interdependent on each other to cast an incremental effect on production of coal.

5. Conclusion

Challenge in mining technologies and management for Indian mining industry is not only to improve efficiency to reduce costs but to have right time and amount set to achieve competitive priorities most efficiently. Even though open cast mining sector is conservative, recent economic crisis has contributed to awareness that there is a wide possibility for managing technology towards effectiveness in terms of cost savings. The proposed study will successfully targeted two basic goals: firstly '*making mine operations easier*' by developing an efficient technology and second by managing in a most effective way. There are some important points should be taken care of it which is mentioned below:

- Installed technology should be properly managed and utilized in order to match the planned and actual output of coal and to increase productivity. To do this, trained employees are needed. It's also crucial to improve technological competitiveness by considering long-term planning, level of management, appropriate technology selection, land acquisition issues, technical skills, proper machine usage, and training requirements that are necessary for mining industry. Since most of the installed technology at Coal India Ltd. and its subsidiaries is imported, effective planning, selection, and implementation of the technology are required. According to the working and geophysical circumstances and factors in India, the design, technical, and operational feasibility should be carefully taken into account.

- The production of the mines would grow as a result of all these factors. To harness all the potential benefits of the installed technology its effective and efficient management is required, a properly planned and managed technology can earn and contribute to organizational performance and success. To manage the change brought on by the introduction of new technologies, effective management must make difficult but doable decisions related to the discovery and appraisal of technologies, the development of new or enhanced procedures, and the integration of technology with other business processes.

- On the other hand, productivity is not result of unit action rather it calls for combination of actions. These actions are required to create an interrelated and interdependent series to bridge the gap between planning and its effective implementation. Pertaining to the research conducted 24 variables were identified, each variable was having their unique effect on the technology installed, their usage and on their effective management. After conducting factor analysis, it was found that 12 variables were statistically significant. These variables are basically the fragments of an optimal concept called management. On the other aspect, it was concluded that target production is more than actual production because different variables (identified during study) effects management of technology and their optimal utilization. Management of installed technology is very much necessary to enhance productivity in open cast mining and after factor analysis is three extraction factors were identified under which these 12 variables were grouped. Management of technology is integration of planning, selection and implementation which is evident in the factor extracted from the variables identified. Overall analysis pinpointing that decreased production as a result of improper management of installed technology. By concentration on strategic, tactical and operational aspect, management of technology can be made efficient and further productivity on coal mining can be enhanced.

Author details

Anand Pd Sinha[1*], Neha Choudhary[1], Rahul Rai[1], Ashok Kumar Asthana[2] and Praveen Chandra Jha[1]

1 Department of Management, BIT Mesra Ranchi, India

2 Sarla Birla University Ranchi, India

*Address all correspondence to: anand.pd.sinha@bitmesra.ac.in

References

[1] Anandarajan I, Akhilesh KB. Effective utilization of tacit knowledge in technology management. In: Driving the Economy through Innovation and Entrepreneurship: Emerging Agenda for Technology Management. India: Springer; 30 Jan 2013. pp. 461-471

[2] Wexler AC, De Loecker J. Impact of Productive Efficiency through Reallocation and Technology - Evidence from US Steel Industry. Centre for Policy Studies. USA: Princeton University; 2013. First Version 2012, Current Version 2013

[3] Bowonder B, Miyake T. Technology management: A knowledge ecology perspective. International Journal of Technology Management. 2000;**19**(7–8): 662-684

[4] David B, Kirit V. Meeting Technology Needs of Enterprises for National Competitiveness. Viyana, Austria: UNIDO; 2001. pp. 7-9

[5] Bright JR. Some management lessons from technological innovation research. Long Range Planning. 1969;**2**(1):36-41

[6] Roy S, Singh SN. Strategic planning for Management of Technology - an empirical study with Indian steel sectors. Published in 'VISION-The Journal of Business Perspective'. 2014;**19**(2):112-131 published by SAGE publications on behalf of MDI-Gurgaon, India

[7] Chaudhuri S, Moulik T. Learning by doing: Technology transfer to an Indian manufacturing firms. Economic and Political Weekly. 1986;**21**(8)

[8] Cattel RB. The scree test for the number of factors, multivariate behavioral research. Multivariate Behavioral Research. 1966;**1**:245-276

[9] Khalil T. Technological competitiveness in the global economy. International Journal of Technology Management. 1992;**7**(5):335-339

[10] Flynn B, Sakakibara S. Empirical research methods in operations management. Journal of Operation Management. 1990;**9**:250-284

[11] Möhring MM. Innovation in a High Technology B2B Context, Exploring Networks, Processes and Management. 2013

[12] Floyd C. Managing Technology for Corporate Success. Aldershot: Gower; 1997

[13] ESCAP U. Technology for development: Study by the ESCAP secretariat for the fortieth session of the commission. 1984

[14] Narayanan VK. Managing Technology and Innovation for Competitive Advantage. India: Pearson Education Inc.; 2001. pp. 76-77

[15] Ford E. Develop your technology strategy. Long Range Planning. 1988;**21**(5):85-95

[16] Watson C. 10 Is preventive maintenance worthwhile? In: Operational Research in Maintenance: Papers Presented at the Symposium on Operational Research in Maintenance Held at Strathclyde University. Barnes & Noble; December 1968. 1970. p. 142

[17] Steele LW. Managing Technology - the Strategic View. New York, USA: McGraw-Hill; 1989

[18] Machado FM. Aspects of Technology Management at the Industrial Enterprise

Level. In: Strengthening Technological Capability. New Delhi: Gyan Publishing House; 1992

[19] Joshi B. Management of Technological Change in the public sector Enterprises in India. In: Management of Technological Change. India: Allied Publishers Limited; 1991

[20] Khalil T. Management of Technology - the Key to Competitiveness and Wealth Creation. New Delhi: Tata McGraw-Hill; 2000. pp. 37-39

[21] Lee J, Lapira E, Bagheri B, Kao H. Recent advances and trends in predictive manufacturing systems in big data environment. Manufacturing Letters. 2013;**1**(1):38-41

[22] Momaya K, Ajitabh A. Technology management and competitiveness: Is there any relationship? International Journal of Technology transfer and Commercialization. 2005;**4**(4): 518-524

[23] Shrouf F, Ordieres J, Miragliotta G. Smart factories in Industry 4.0: A review of the concept and of energy management approached in production based on the Internet of Things paradigm. In: 2014 IEEE International Conference on Industrial Engineering and Engineering Management. IEEE; 2014. pp. 697-701

[24] Natarajan R. The nature and scope of technology management. In: Proceedings to National Seminar on Technology Management. Indian National Academy of Engineering; 1999. pp. 33-42

[25] Iarovyi S, Mohammed WM, Lobov A, Ferrer BR, Lastra JLM. Cyber–physical systems for open-knowledge-driven manufacturing execution systems. Proceedings of the IEEE. 2016; **104**(5):1142-1154

[26] Millett Stephen M. The strategic Management of Technological R&D. International Journal of Technology Management. 1990;**5**:153-163

[27] Burgelman RA, Maidique A, Wheelwright SC. Strategic Management of Technology and Innovation. New York, USA: McGraw-Hill; 2001

[28] Dodgson M, Gann D, Salter A. The Management Technological Innovation Strategy and Practice. UK: Oxford University press; 2008

[29] Rastogi R, Dave V. Managerial effectiveness: A function of personality type and organizational components. Singapore Management Review. 2004; **26**(2):79-87

[30] Nawaz S. Management of Technology Transfer and Development. Bangalore: APCTT; 1983

[31] Anders D, Soren G, Holm C. Management of Technology in a complex world. International Journal of Materials and Product Technology. 1997; **12**(4–6):239-259

[32] Flippo EB. Personnel Management McGraw–Hill. International Book Company; 1980

[33] Ghatak M, Jiang NN. A simple model of inequality, occupational choice, and development. Journal of Development Economics. 2002;**69**(1):205-226

[34] Bancroft NH, Mumford E. New partnerships for managing technological change. New York, NY: Wiley; 1992

[35] Saxena SC, Jotshi CK. Management and combustion of hazardous wastes. Progress in Energy and Combustion Science. 1996;**22**(5):401-425

[36] Boskin MJ, Lau LJ. Capital, technology, and economic growth. In: Boskin MJ, Lau LJ, editors. Technology and the Wealth of Nations. Vol. 17. 1992

[37] Khalil T. Management of Technology-The Key to Competitiveness and Wealth Creation. 3rd ed. New Delhi: Tata McGraw-Hill; 2010. pp. 4-11

[38] Christina TB. Effective management of technology implementation, Journal of Healthcare Management. California, USA; 2009

[39] Ghatak S. A Study of Mine Management Legislation and General Safety. 13th ed. Asansol: Coalfields Publishers; 2009

[40] Solomon JJ. Importance of Technology Management for Economics Development. International Journal of Technology Management. 1990;**5**(5): 523-536

[41] Khalil T. Management of Technology-The Key to Competitiveness and Wealth Creation. third ed. New Delhi: Tata McGraw-Hill; 2009. pp. 4-7

[42] Betz F. Managing Technology. Englewood Cliffs, NJ, USA: Prentice Hall; 1987

[43] Mehta MM. Measurement of Industrial Productivity. 2002. p. 31

[44] Maria M. Innovation in a High Technology B2B Context: Exploring Supply Networks Processes and Management. Springer Verlag; 2014

[45] Moustafa ME. Management of Technology Transfer. Geneva, Switzerland: International Labour Organization; 1990 Newsletter of ENVIS Nodal Centre on Environmental Problems of Mining Areas, Number 24 & 25

[46] Vaux A, Riedel S, Stewart D. Modes of social support: The social support behaviors (SS-B) scale. American Journal of Community Psychology. 1987;**15**(2): 209-232

[47] Stewart F. Macro-Policies for Appropriate Technology in Developing Countries. USA: Westview Press, Boulder, Co.; 1987

[48] Sevim H, Lei DD. The problem of production planning in open pit mines. INFOR: Information Systems and Operational Research. 1998;**36**(1-2):1-2

[49] Ritson J. The Health and Safety at Work Act. Ravenswood Publications Ltd; 1983

[50] Pal K, Verma HL. Foundation of Management. New Delhi: Vayu Education of India; 2009. pp. 10-14

[51] Koontz TM, Bodine J. Implementing ecosystem management in public agencies: Lessons from the US Bureau of Land Management and the Forest Service. Conservation Biology. 2008;**22** (1):60-69

[52] Singh G. Regional Environmental and Social Challenges Facing the Coal Industry. Indian Coal Forum and Mgmi: The Coal Summit Organized the India Energy Forum; 2005. pp. 19-20

[53] Chhipa MK, Kumar G. Year of Publication: 2014. Technology Letters. 1999;**11**:691-693

[54] Berman EM. Technological competitiveness in the global economy. International Journal of Technology Management. 1992;**7**(5):347-358

[55] Sun H. Pattern of Organizational and Technological Development with Strategic Consideration. Aalberg, Denmark: Aalberg University; 1993

[56] Mashelkar DR. Five Technology Management Mantras for Indian Industry. In: Proceedings of the 2001 National Seminar on Technology Management. Indian National Academy of Engineering; 2001

[57] Gregory MJ. Technology management - a process approach. Proceedings of the Institution of Mechanical Engineers. 1995;**20**(9):347-356

[58] Schimm B. Under the surface. World Coal (Dorking). 2004:13

[59] Ulhoi J. Towards theoretical and methodological corporate technology management framework-the strategic perspective. International Journal of Management. 1996;**12**(2):199-209

[60] Lan PMC. Technology management - a complex adaptive systems approach. International Journal of Technology Management. 2003;**25**:728-745

[61] Bhalla SK. The Effective Management of Technology. Columbus OH, USA: Battelle Press; 1987

[62] Besterfield DH, Besterfield-Michna C, Besterfield-Sacre M, Besterfield GH, Urdhwareshe H. Total Quality Management. For Anna University. Pearson Education India; 2011

[63] Vanneste SG, Van Wassenhove LN. An integrated and structured approach to improve maintenance. European Journal of Operational Research. 1995;**82**(2):241-257

[64] Perrino AC, Tipping JW. Global Management of Technology. Research Technology Management. 1989;**32**(3):12-19

Chapter 4

Analysis of Quality Control of the Production Process of Rotary Kiln III Using the Lean Six Sigma Method at PT. XYZ Southeast Sulawesi

Ahmad Padhil, Nurhayati Rauf and Ayu Reski Ilahi

Abstract

Quality control is a verification system from a process level to measure the characteristics of product quality, compare specs and ensure quality in accordance with the predetermined standards. PT. XYZ Southeast Sulawesi is one of the FeNi (FerroNickel) processing factory companies that really pay attention to the quality of the products produced, in the work division at PT. XYZ Southeast Sulawesi still has defective products, one of which is the Rotary Kiln III section. Therefore, research is carried out in that section to detect out what factors cause product failures and seek improvement suggestions to minimize the types of failures that occur. The method used is Lean Six Sigma. The stages in this analysis use the define, measure, analyze, improve, and control (DMAIC) stage. The average value of the Sigma level in the Rotary Kiln III section is 4028 with a DPMO value of 5745. In the Rotary Kiln III section, the root cause analysis of the problem with a Fishbone diagram is then executed to improve using the FMEA method.

Keywords: quality control, Lean Six Sigma, DMAIC, fishbone, FMEA, clinker, RPN

1. Introduction

Quality is a dynamic condition related to products, services, people, processes and the environment that satisfy or exceed consumer expectations [1]. Good quality according to the manufacturer is if the product produced by the company is in accordance with the specifications determined by the company, continuous quality improvement is absolutely necessary in industrial competition [2]. Product quality is an overall evaluation process to customers for improving the performance of a product [3]. Quality control is a system of verification and maintenance or maintenance of a desired level or process by means of careful planning, use appropriate equipment, continuous inspection, and corrective action where necessary [4, 5]. Quality control is an activity (company management) to maintain and direct the quality of the company's products or services as planned by Ahyari [6]. Minimizing defects is an effort that must be performed continuously in terms of improving the quality of a product. Therefore, it is very important for companies to implement a method of quality control and improvement that can help reduce defects in developing products [7].

PT. XYZ Southeast Sulawesi is one of the companies processing FeNi (FerroNikel). The processes that exist in PT. XYZ Southeast Sulawesi starts with ore preparation, smelting, to refining. One of the ferronickel processing processes at PT. XYZ Southeast Sulawesi is the calculation process of laterite nickel ore in Rotary Kiln. The calcination process is a part of ore preparation, which aims to prepare the laterite nickel ore before smelting, namely by suppressing the water content of the crystals in the ore while reducing some of the ore to metal. The calcination process in the Rotary Kiln often feels so much that the Clinker is formed.

In the Rotary Kiln production process, at that place are operating parameters that must be considered in order for the process to run smoothly, the first parameter is the fullness of the Rotary Kiln, fullness, which is the number of ores that fill the kiln space. The next parameter is the operating temperature, if the operating temperature is too low, the LOI level will be high, that is, when the LOI is >1%, thus reducing the quality of calcine, if the operating temperature is too high it will increase the possibility of clinker formation. The next parameter is the retention time duration, if the retention time duration is also too low, the heat received by the ore will not be evenly distributed so that the moisture content of the ore is not reduced maximally, if the retention time duration is too fast, the potential for clinker will increase and result in a lack of calcine production.

PT. ANTAM Tbk. Southeast Sulawesi UBPN Rotary Kiln III department has a production target of Condition Ore of 60,140 tons or 42,101 tons of calcine (**Figure 1**).

Six sigma is a method that is being developed in today's world. The application of six sigma is expected to reduce failure (damage) in achieving the desired quality goals in increasing the amount of production [8]. Lean Six Sigma is an interesting method used to measure quality and make improvements to improve the quality of goods or services [9]. Lean and Six Sigma integration will improve business and industry performance through increased speed and accuracy [10].

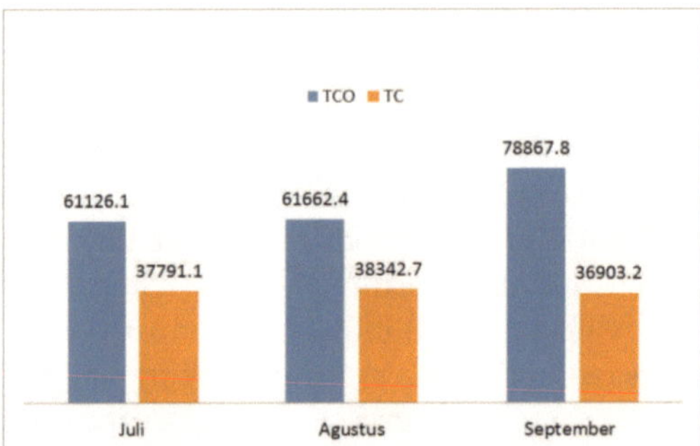

Figure 1.
Production result TCO & TC.

2. Research methodology

2.1 Time and place of research

The place of research in this writing was conducted at PT. XYZ Southeast Sulawesi on Jalan Jend. Ahmad Yani No. 5, Pomalaa, Kolaka Regency, Southeast Sulawesi. The research was conducted for approximately 1 month start from July – August.

2.2 Data sources

1. Primary data are in the form of production data and breakdown data.

2.3 Data collection methods

2.3.1 Observation

Through this observation technique the author collected data by making direct observations at the Rotary Kiln III department.

2.3.2 Interview

A method for receiving data and information by communicating directly with Rotary Kiln III assistant manager.

3. Results and discussion

3.1 Define

This stage contains data on the flow of the production process starting from the Supplier to the Customer in the Rotary Kiln III section and identifies what wastes are in the Rotary Kiln III section.

From the **Figure 2** above, it explains that the supplier of the Rotary Kiln section is Ore preparation, the input is in the form of ore preparation, the Rotary Dryer input is ore with MC 22 ± 1% and the input ore mixes is an ore + coal condition which will be processed in the production process. The production process is carried out by the drying zone section with a temperature of 250–300°C for 47 minutes, Preheating zone

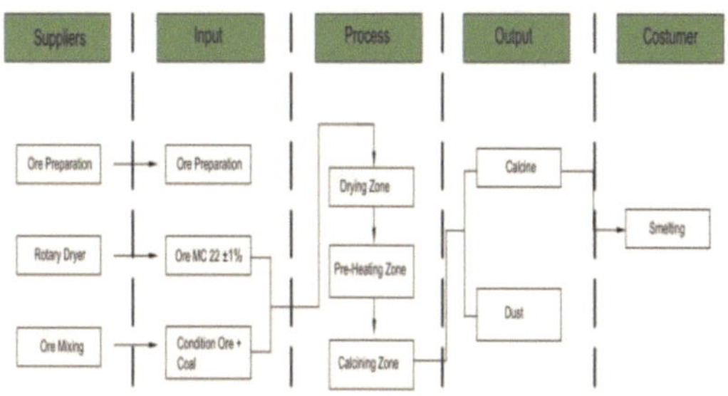

Figure 2.
SIPOC diagram.

No	Month	Sample Rotary Kiln III	Number of Defects	Proportion of Defects	UCL	CL	LCL
1	July	61126.1	1055.2	0.01726268	0.241962411	0.017123	−0.207437061
2	August	61662.4	1142.6	0.01852993	0.243229667	0.017123	−0.206169805
3	September	78867.8	1255.2	0.01591524	0.240614977	0.017123	−0.208784496
	Total	201656.3	3453	0.051707846			

Table 1.
Control chart calculation.

with a temperature of 700–850°C for 47 minutes, and a calcining zone with a temperature of 900–1000°C for 35 minutes. This section removes moisture until the water content is below 1%. Furthermore, the output of the process is in the form of calcine and dust where the calcine will be processed to the next stage, namely the Smelting department. The zones that produce the most defective products are the preheating zone and the calcining zone. Defective products in this section can come from suppliers and originate from the production process itself (**Table 1**).

3.2 Measure

3.2.1 Determination of product control limits

1. Calculating Proportion of Defects

$$P = \frac{\mathrm{np}}{\mathrm{p}} \tag{1}$$

2. Calculating the Center Line (CL)

$$\mathrm{CL} = \dot{\mathrm{p}} \tag{2}$$

3. Calculating Upper Control Limit (UCL)

$$\mathrm{UCL} = \dot{\mathrm{p}} + 3\left(\sqrt{(\dot{\mathrm{p}}\,(1 - \dot{\mathrm{p}})/\mathrm{n}}\right. \tag{3}$$

4. Calculating Lower Control Limit (LCL)

$$\mathrm{LCL} = \dot{\mathrm{p}} - 3\left(\sqrt{(\dot{\mathrm{p}}\,(1 - \dot{\mathrm{p}}))}\right)/\mathrm{n} \tag{4}$$

3.2.2 Calculation of DPMO (defect per million opportunity) & sigma level

The results of the calculation of the DPMO value are used to determine the ratio of defects one per one million opportunities. From the calculation results, the average DPMO value of 5745 means that there is a possibility of 5745 defects that will occur in

Defect	CTQ	DPU	TOP	DPO	DPMO	Sigma
1055.2	3	0.017	183378.3	0.006	5754	4.027
1142.6	3	0.019	184987.2	0.006	6177	4.002
1255.2	3	0.016	236603.4	0.005	5305	4.055
3453		Average			5745	4.028

Table 2.
DPMO Value & Sigma level value.

one million outputs or units of Rotary Kiln III resulting from the Drying Zone process, Preheating Zone, and Calcining Zone. Meanwhile, if converted into sigma value, the value obtained is 4028, which is the achievement of the industry average six sigma level (**Table 2**).

3.3 Analyze

The use of the Fishbone diagram is to see the relationship between the problems faced with the possible causes and the factors that influence it. The Fishbone diagram is an analytical tool used to analyze what happens in the production process resulting in the formation of clinkers (**Figure 3**).

The following is a discussion of the Fishbone diagram and validation of the causes of defects of each type of defect that occurred in the Rotary Kiln III section:

4. Method

The absence of an appropriate temperature standard for each production resulted in the formation of excessive clinker.

4.1 Material

a. A full top bin result in stopping the production process to wait for the top bin to be empty or ready to be filled.

b. The exhaustion of fuel results in the cessation of the production process because the production process is not optimal.

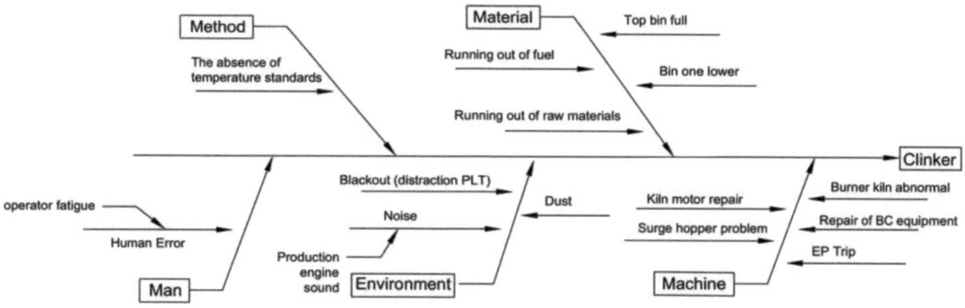

Figure 3.
Fishbone diagram.

c. Bin ore lower occurs due to a decrease in the production process in ore mixing, which result in a halt in the production process at Rotary Kiln III to wait for bin ore to be refilled

d. Raw Material depletion is very rare but this occurs when production decreases or there is a problem with the Rotary Dryer which result in the production process of ore mixing not being carried out.

4.2 Man

Human Error that occurs when the operator is tired and becomes unfocused and causes the error to increase or decrease the temperature during the production process and also the rotation duration of the Rotary Kiln.

4.3 Environment

a. Blackout Blackout (PLT disruption) resulted in the cessation of all production processes that are being carried out in each existing department. This usually happens when there is interference that occurs in the PLT itself.

b. The dust that is generated around the production floor interferes with employee activities at work, because it sometimes interferes with breathing.

c. Noise due to the sound of production machines greatly disturbs the focus of employees in carrying out production activities.

4.4 Machine

a. Abnormal Kiln Burner results in inappropriate combustion fuel and air which results in stopping the production process for checking.

b. Repair of the Kiln Motor is carried out when the kiln rotation does not rotate as specified in the setting.

c. Repair of Belt Conveyor (BC) Equipment resulted in the cessation of material (ore) transportation into the kiln room.

d. Problematic Surge Hopper resulted in the interruption of the production process because checks and repairs had to be done in order to accommodate the production of Rotary Kiln III.

e. Electrostatic Partikel (EP) Trip causes fine dust on the Rotary Kiln not to be caught and enter the production results.

4.5 Improve

Failure Mode and Effect Analysis (FMEA) is used to determine the priority level of the causes of defects that occur [11]. From the Risk Priority Number value obtained from the Severity, Occurrence and Detectability values, it shows that the causes that have the highest RPN value can be made improvements to reduce or even eliminate these defects.

Failure Mode and Effect Analysis (FMEA) is used to determine the priority level of the causes of defects that occur. From the Risk Priority Number value obtained from the Severity, Occurrence and Detectability values, it shows that the causes that have the highest RPN value can be made improvements to reduce or even eliminate these defects (**Table 3**).

The highest RPN value with the cause of the resulting combustion heat is not in accordance with the RPN value of 448. Suggestions for improvements that can be made are scheduling maintenance and scale inspection.

The lowest RPN value caused the disruption of employee activities with an RPN value of 60. Suggestions for improvements that can be made are adding or updating EP.

4.6 Control

For quality control proposals, namely continuous improvement to reduce defective products that arise so that production targets can be increased and as expected. The

Failure Mode	Cause of Failure	Proposed Improvement	RPN
Clinker	The resulting combustion heat is not suitable	Maintenance scheduling and periodic inspections	448
	Lack of checking during maintenance	Maintenance scheduling and periodic inspections	392
	Lack of maintenance	Maintenance scheduling and periodic inspections	343
	Conveyor Belt transports excess material	Maintenance scheduling and periodic inspections	294
	There is no suitable scheduling	Make a schedule to keep the production process running	240
	Incorrect machine settings	Perform checks and set standards for temperature	224
	The amount of dust generated during production	Maintenance scheduling and periodic inspections	210
	Lack of proper checking and scheduling	Set a schedule for periodic checks	200
	Lack of checking	Set a schedule for checking	180
	Poor equipment performance and interruption of the operation process	Inspection of tools that support engine performance at PLT	160
	Lack of scheduling checks	Set a schedule for checking	150
	Reduces employee focus	Implement a sound suppressor	84
	Employees lack focus	Monitor operator performance	70
	Interferes with employee activities	Add or update EP	60

Table 3.
Proposed repair and sequence of causes of failure based on RPN.

implementation of lean six sigma in the company can increase the current sigma value of the company so that the company can strive to achieve a 6 sigma value. The following are proposed controls that can be used to address the root causes of existing problems:

a. Check and clean the machine regularly

b. Creating a check form as a control of machine conditions.

c. Perform periodic machine condition reports.

5. Conclusions and suggestions

5.1 Conclusion

The conclusions obtained after processing and analyzing data are as follows:

1. The average value of DPMO is 5745 with a sigma value of 4028 out of 2570, which means that the capability is good enough, but it is necessary to control the quality so that the resulting product reaches zero defects.

2. To minimize failures that occur, it is necessary to schedule machine maintenance and periodic checks. And by knowing the factors that cause the failure that occurs, the quality will increase so that the production target will be achieved. The most influential factors causing the failure are the Abnormal Kiln Burner, the repair of the kiln motor, and the problematic Surge Hopper.

5.2 Suggestion

Suggestions that can be given to companies to become input for the Rotary Kiln III section in an effort to reduce failed products and control the production process are as follows:

1. By knowing the sigma level on product failure at Rotary Kiln III, it is expected that the company can minimize failed products so that the company's sigma level value can increase and the resulting failed products can decrease.

2. We recommend that the company be able to periodically check machines with a planned schedule so that the production process can be controlled and the expected production targets can be achieved.

Acknowledgements

Contain the appreciation given by the author to those who have played a role in the research, both in the form of funding support, licensing, consultants, and assisting in data collection. Especially to the Department of Industrial Engineering, Muslim University of Indonesia and the IntechOpen book chapter for the opportunity.

Author details

Ahmad Padhil*, Nurhayati Rauf and Ayu Reski Ilahi
Departement Industrial Engineering, Universitas Muslim Indonesia, Makassar, Indonesia

*Address all correspondence to: ahmad.padhil@umi.ac.id

References

[1] Santa Hadi. Kualitas Produk vs Kepuasan Pelanggan [Internet]. 2015. Available from: http://kompasiana.com/hadisanta/kualitasprodukvskepuasanpelanggan [Accessed: Desember 28, 2019]

[2] Wahyu AD. Manajemen Kualitas. Yogyakarta: Andy Offset; 2009

[3] Mowen JC, Minor DM. Perilaku Konsumen dialih bahasakan oleh Dwi Kartika Yahya. Jakarta: Erlangga. PT Antam Tbk., FeNi Smelting Plant No. 3, Pomalaa: Process Descriptions. 2003; 2012

[4] Padhil A, Pawennari A, Dahlan M, Awaliah NR. Usulan Perbaikan Lingkungan Kerja pada Bagian Mesin Puffing Gun di IKM Bipang Putri Sehati Kabupaten Gowa. Journal of Industrial Engineering Management. 2018;**3**(1):1-9

[5] Sulaeman. Analisa Pengendalian Kualitas Untuk Mengurangi Produk Cacat Speedometer Mobil dengan Menggunakan Metode QCC di PT. INS. Jurnal Pasti. 2014;**VIII**(1):71-95

[6] Ahyari A. Manajemen Produksi. Jilid 1. Yogyakarta: Erlangga Offset; 1992

[7] Kartika H. Analisis Pengendalian Kualitas Produk CPE film Dengan Metode statistical process control Pada PT. MSI. Jurnal Ilmiah Teknik Industri Tahun 2013. 2013;**1**(1):50-58

[8] Didiharyono DKK. Analisis Pengendalian Kualitas Produksi dengan Metode Gaspresz, Vincent dan Avanti Fontana. In: Lean Six Sigma for Manufacturing and Service Industries. Bogor: Penerbit Vinchristo Publication; 2018

[9] Budiawati H. Pendekatan Lean Six Sigma Dalam Penentuan Prioritas Perbaikan Layanan Bank Berdasarkan Persepsi, Harapan dan Kepentingan Nasabah. Journal Management. 2017; **XXI**(1):1-6

[10] Didiharyono D, Marsal M, Bakhtiar B. Analysis of production quality control with six-sigma method in the drinking water industry of PT Asera Tirta Posidonia, Palopo City. Sainsmat. Scientific Journal of Natural Sciences. 2018;7(2):163

[11] Padhil A, Purnomo H. Macroergonomic approaches as a solution to local wisdom-based tourist village development planning. In: MATEC Web of Conferences 2018. Vol. 154. EDP Sciences; 2018

Section 3

Application Technologies

Chapter 5

Fluidization Behavior of Binary Mixtures of Coal in a Top-Fed Bubbling Fluidized Bed Gasifier

Ali Can Sivri

Abstract

Bubbling Fluidized Bed Gasifier (BFBG) technology is an efficient and economical way of producing syngas from various feedstocks, such as coal, biomass, and municipal waste. However, the prediction of the gasification process inside the BFBG is quite complex due to many factors, including multiphase flow hydrodynamics. This study analyzed the hydrodynamics of a bench-scale top-fed bubbling fluidized bed coal gasifier with sand or glass beads used as bed materials at different bed aspect ratios. Two separate test rigs were built with the same dimensions for cold flow (without reaction) and hot flow (with reaction) studies, respectively. The cold flow test rig was used to investigate the hydrodynamics of BFBG fluidization. Bed pressure drop, minimum fluidization velocity, and mixing were analyzed in the test room conditions. Following that, gasification tests were carried out in the hot flow BFBG test rig with a novel feeding system using the optimum hydrodynamical parameters determined from cold flow analyses. Results showed that syngas was successfully produced at an adequate composition. This study contributes to a better understanding of the fluidization hydrodynamics of the binary coal and bed material mixtures in a top-fed BFBG for a more optimum gasification process and easier operation of the BFBG.

Keywords: bubbling fluidized bed gasifier, coal fluidization, coal gasification, fluidization hydrodynamics, multiphase flow, synthetic gas

1. Introduction

The demand for energy has been growing due to the increase in human population and industrialization. However, the greater need for fuel and power generation brings more greenhouse gases and, hence, more environmental pollution, mostly because of the dominance of fossil energy sources being used as primary fuels for transportation and power generation. According to the International Energy Agency (IEA) and British Petroleum (BP), coal was the primary energy source, accounting for 38.5% of electricity production from 1971 to 2021 and 36% in 2021. Gasification can be an environmentally friendly alternative way of generating fuel (synthetic gas, syngas) from coal. Syngas produced by coal gasification can be used in various applications, such as transportation fuel [1], electricity production, heating, etc. Besides, the use of syngas can significantly lower greenhouse gas emissions. Bubbling fluidized bed gasifiers are a type of fluidized bed reactor that can significantly increase the gasification reaction efficiency compared to fixed-bed reactors. Also, it is more economical to operate and maintain compared to Circulating Fluidized Beds (CFB). However, the BFBG fluidization process is quite complex, and it has a direct impact on the reaction process and its efficiency. There are many parameters affecting the fluidization hydrodynamics inside the BFBG. Particle characteristics such as size, sphericity, density [2, 3], bed aspect ratio (the ratio of the bed height to the bed diameter) [3, 4], fluidizing gas and feedstock moisture content [5], temperature, and feeding location have strong effects on fluidization hydrodynamics. Besides, fluidizing gas velocity is a crucial parameter that mainly regulates and determines the fluidizing regime according to the particle characteristics and other factors. For an efficient gasification reaction inside a BFBG, all these parameters should be considered and adjusted interactively.

Particle size and density affect the bubble-induced particle mixing in a gas–solid fluidized bed. Because, particle shape affects the bubble formation and dynamics in fluidized beds. Spherical particles tend to form small and uniform bubbles, while non-spherical particles tend to form large and irregular bubbles [6]. Larger and denser particles tend to mix slower than smaller and lighter particles. Si and Guo [3] studied the fluidization behavior of binary mixtures of quartz sand with sawdust or wheat stalk in an acoustic bubbling fluidized bed. They found that the addition of sand improved the fluidization quality of the biomass particles. The authors suggested that the improved fluidization behavior was due to the increased particle density and reduced voidage caused by the addition of sand. They also observed that the fluidization behavior of the binary mixtures was affected by the particle size ratio and the proportion of sand in the mixture.

Higher aspect ratio beds have higher reaction efficiencies due to increased interparticle attraction and gas residence time, resulting in a longer reaction time. However, they lead to poorer mixing due to the transition from single bubble regime to slug flow regime [7]. Feedstock particles need to be delivered homogeneously along the bed height for optimum particle interaction and heat transfer rates. Thus, homogeneous mixing is suggested to obtain better syngas composition and higher reaction efficiencies. Most of the applications deliver the feedstock particles into the bed mainly by pushing the feedstock directly into the bed or on top of the bed by using a screw driver. However, driving a screwdriver directly into the reactor bed through the reactor wall can be expensive, and it is prone to mechanical and operational problems such as leaks. Zhijie Fu et al. [8] studied the particle mixing and segregation behavior in an Air Dense Medium Fluidized Bed (ADMFB) with binary mixtures of solid particles for dry coal beneficiation. The study examined the effects of various operating parameters such as particle density ratio, particle size ratio, mixture composition, superficial gas velocity, and fluidized bed height on the mixing and segregation pattern. The results of the study show that the degree of segregation increases with increasing density difference of binary mixtures and partial segregation can occur with an increase in particle size ratio. However, the mixing and segregation of binary systems are almost independent of lower excess gas velocity and initial bed height when it is over 15 cm. The study also employs a mixing index to evaluate the mixing and segregation performance and identifies criteria for good mixing to achieve the bed density adjustment.

During the operation of the BFBG, bed pressure drop and minimum fluidization velocity are the two main parameters analyzed to control the fluidization behavior inside the fluidized bed. Particle density affects the minimum fluidization velocity and bed expansion of fluidized beds. Gao et al. [9] showed that higher particle density leads to higher minimum fluidization velocity and lower bed expansion. Abdullah et al. [10] investigated the effect of mixture bulk density and bed voidage on the minimum fluidization velocity (U_{mf}) of various materials, including Geldart B-group materials. They found that both mixture bulk density and bed voidage had a significant effect on the minimum fluidization velocity of the materials. In addition, they found that Geldart B-group materials exhibited better fluidization behavior compared to other groups of materials in Geldart's classification. On the other hand, many studies, conducted to analyze the effect of the bed aspect ratio on the U_{mf}, show that higher bed aspect ratios do not affect the U_{mf} significantly [11, 12]. Many predictions have been made to predict the U_{mf}, but many of them do not agree, particularly for the binary mixtures (feedstock and inert materials). Besides, studies made in lab-scale applications can generate results that differ significantly in real (larger-sized) applications. In addition, most of these correlations have been generated in ambient conditions, excluding the temperature effect on fluidization hydrodynamics [13].

Temperature and pressure affect the hydrodynamics of gas–solid fluidized beds by changing the gas properties (density and viscosity) and the interparticle forces (van der Waals, electrostatic, etc.). Higher temperature and pressure increase the interparticle forces and cause agglomeration and defluidization of fine particles. The fluidization behavior of different Geldart groups of particles (A, B, and D) can vary significantly under extreme conditions [14]. Higher temperature leads to lower gas density and higher gas viscosity, which decrease the minimum fluidization velocity and increase the drag force [14, 15].

Overall, the mixing characteristics and quality of binary mixtures in fluidized beds can be an important aspect to consider in various industrial applications such as coal combustion, gasification, and catalytic reactions. According to the literature review, the theory behind fluidization hydrodynamics is not fully developed and understood. Hence, further research is needed to fully understand and optimize the mixing behavior of these complex mixtures under different operating conditions. Few studies considering the temperature effect on the fluidization characteristics and mixing behavior of binary mixtures of coal and inert materials in a fluidized bed have been found in the literature. As a result of the complexity of the fluidization and gasification theories, as well as the difficulties in their application, this technology is not widely used in both large- and small-scale applications.

This study aims to contribute to a better understanding of the coal and inert material (glass beads or sand) fluidization behavior in a top-fed deep-bed application of a BFBG by considering the particle characteristics, temperature, and bed aspect ratio.

1.1 Gasification and bubbling fluidized bed gasifier

Gasification is also known as "partial combustion" or "oxidation" due to the lower oxygen requirement (25 to 40%) compared to the amount of oxygen used in the stoichiometric combustion reaction. As a result, in addition to syngas, gasification generates some carbonaceous by-products such as ash, tar, and char (**Figure 1**). A series of reactions that take place interactively in a gasification reaction are shown in **Table 1**.

A bubbling fluidized bed gasifier is a type of Fluidized Bed Reactor (FBR) in which gasification takes place with a complex multiphase fluidization process of gas and

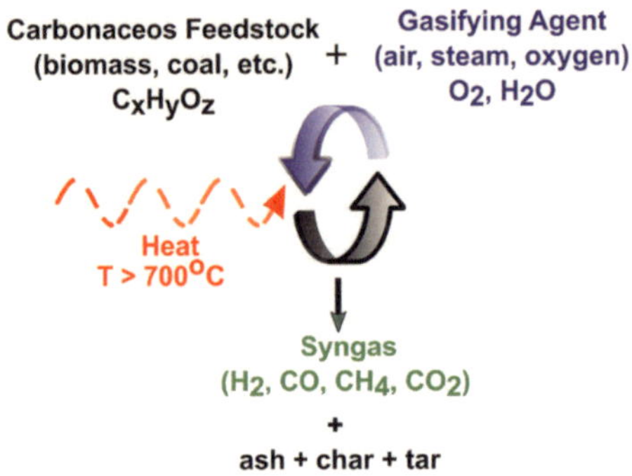

Figure 1.
Gasification reaction diagram.

Reaction name	Reaction formula	Enthalpy values
Water-gas shift reaction	$CO + H_2O \rightleftharpoons CO_2 + H_2$	$\Delta H^\circ 298K$ = -41 $kJmol^{-1}$
Steam methane reforming reaction	$CH_4 + H_2O \rightleftharpoons CO + 3H_2$	$\Delta H^\circ 298K$ = 206 $kJmol^{-1}$
Boudouard reaction	$C + CO_2 \rightleftharpoons 2CO$	$\Delta H^\circ 298K$ = 171 $kJmol^{-1}$
Methanation reaction	$C + 2H_2 \rightleftharpoons CH_4$	$\Delta H^\circ 298K$ = -75 $kJmol^{-1}$
Water-gas reaction	$C + H_2O \rightleftharpoons CO + H_2$	$\Delta H^\circ 298K$ = 131 $kJmol^{-1}$

Table 1.
Major 1st step gasification reactions.

solid particles. Fluidized bed reactors are more efficient in terms of heat transfer and carbon conversion rates compared to fixed-bed reactors, in which there is no particle motion or mixing [16, 17]. Fluidization can be described as the state of the solid particles that are suspended or gained motion (like fluidized) with the lift force exerted by an upcoming fluid (fluidizing fluid or agent) such as gas or liquid in a vertical column. If the flow rate of the gasifying agent is not enough for the desired fluidization operation, additional inert fluidizing gas can be used to obtain the required fluidizing gas flow rates. The mixing and the heat transfer rates are directly related to the bubble dynamics inside the BFBG. Here, besides the bubble dynamics, the inert (bed) material plays a significant role in providing the necessary heat transfer rates to the feedstock particles. Hence, inert material characteristics such as diameter, sphericity, density, and thermal conductivity affect fluidization hydrodynamics, and therefore, heat transfer rates and reaction rates are affected. The term "bed" is used to refer to the mixture inside the reactor. A sample fluidized bed reactor and 3D CAD model illustration are shown in **Figure 2a, b**, respectively. The main parts of the BFBG are: a plenum for fluidizing agent intake, a distributor plate to distribute the flow uniformly above it and provide the bed pressure drop for bubble formation, the reactor bed where the reaction takes place, and a freeboard to decrease the gas velocity to allow the particles to fall back to the reactor bed.

The operation of the BFBG can be described by the following procedures: Typically, a screw feeder mechanism is used to supply feedstock particles to the reactor bed from the bottom, side, or top. The gasification reaction starts when the carbonaceous particles are added to the oxygen-rich atmosphere at the required reaction temperatures. During the gasification reaction with the fluidization process, the

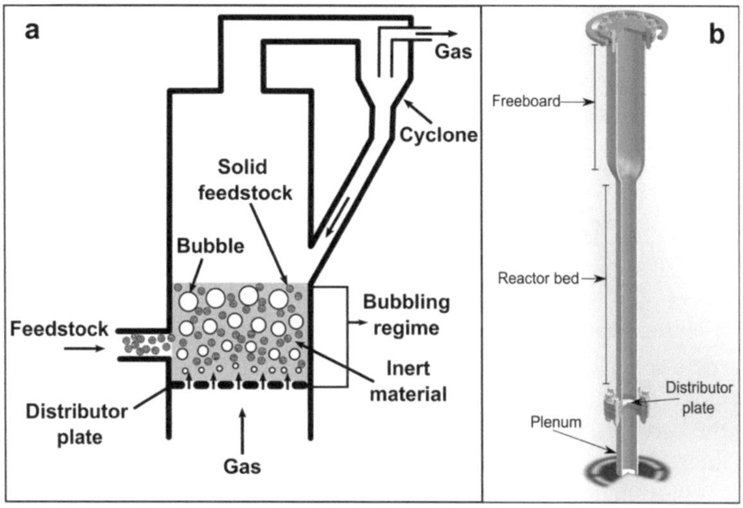

Figure 2.
Fluidized bed reactor diagram.

produced ash particles sink to the bottom of the bed. On the contrary, light particles such as char can leave the reactor at high gas velocities. To increase the gasification reaction efficiency, particularly in circulating fluidized bed (CFB) applications, a cyclone is used to transfer the leaving particles back to the reactor bed. The operation of the BFBGs is more economical compared to CFB applications. Besides, BFBG can use a wider range of materials as feedstocks [18]. Before being used in the different applications previously described, the syngas is cooled and filtered.

1.2 Bed pressure drop and minimum fluidization velocity

The pressure drop between any points along the reactor bed height is equal to the weight of the particles between the measurement points per unit cross-sectional area of the fluidized bed. Thus, the bed pressure drop term is used for the pressure drop of the whole bed weight at the fluidization state. The bed pressure drop can be calculated by using the formula:

$$\Delta P_b A = W, \text{where } W = mg = AH_{mf}\left(1 - \varepsilon_{mf}\right)\left(\rho_p - \rho_g\right)g \tag{1}$$

where, ΔP_b is the bed pressure drop, A is the cross-sectional area, m is the mass of the bed, g is the gravitational acceleration, H_{mf} is the bed height at minimum fluidization, ε_{mf} is the bed voidage at minimum fluidization, ρ_p is the particle density, and ρ_g is the fluidizing gas density. The bed voidage, ε, is the ratio of the void volume to the bulk volume of the bed, can be calculated as:

$$\varepsilon_1 = 1 - \frac{\rho_b}{\rho_s} \tag{2}$$

where ρ_b is the bulk density, and ρ_s is the bed skeletal density.

The minimum fluidization velocity, U_{mf}, is the gas velocity required to balance the bed weight and initiate fluidization. And, theoretically, as the gas velocity increases, the pressure drop per bed weight remains constant. Many correlations have been developed to predict the minimum fluidization velocity; however, most of these correlations, particularly for binary mixtures, do not agree on the prediction results [19]. One of the well-known correlations to predict the minimum fluidization velocity derived by Ergun [20] is:

$$\frac{\rho_g\left(\rho_p-\rho_g\right)\left(gd_p^3\right)}{\mu_g^2}=\frac{150\left(1-\varepsilon_{mf}^2\right)}{\phi^2\varepsilon_{mf}^3}\frac{\rho_g U_{mf} d_p}{\mu_g}+\frac{1.75}{\phi\varepsilon_{mf}^3}\frac{\rho_g^2 U_{mf}^2 d_p^2}{\mu_g^2} \tag{3}$$

where d_p is the mean particle diameter, μ_g is fluidizing gas viscosity, and ϕ is the average particle sphericity.

Furthermore, the minimum fluidization velocity can be determined graphically by measuring the bed pressure drop as the fluidizing gas velocity increases. The intersection of the lines of the slopes in the fixed-bed state and the complete fluidization state, respectively, gives the minimum fluidization velocity. Fluidization starts at the initial fluidization velocity and reaches a complete fluidization state at the complete fluidization velocity. The graphical illustration of the determination of the initial (U_{if}), minimum (U_{mf}), and complete (U_{cf}) fluidization velocities is illustrated in **Figure 3**.

1.3 Fluidization regimes

The potential fluidization regimes within a fluidized bed with increased fluidizing gas velocity (U_g) are shown in **Figure 4**. The bed retains its shape and bulk density as long as the gas velocity remains below the minimum fluidization velocity (U_{mf}). This regime of fluidization is called "packed bed" or "fixed bed," which is illustrated in **Figure 4a**. With the rise in the gas velocity, initially smaller particles in diameter are suspended by the upcoming fluidizing gas. Later, with an adequate flow rate, all particles become lifted, and the weight of the bed is balanced with the force exerted by the fluidizing gas on the bed's cross-section area. The minimum fluidization velocity, U_{mf}, is the gas velocity required to balance the bed weight and initiate fluidization.

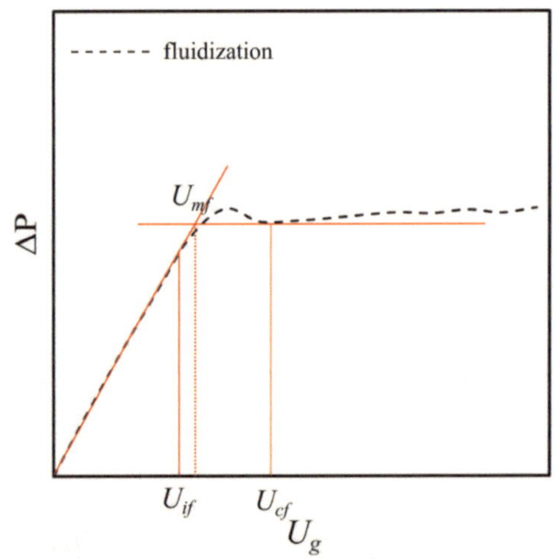

Figure 3.
Graphical solution to determine U_{if}, U_{mf}, and U_{cf} for increasing superficial gas velocity (fluidization).

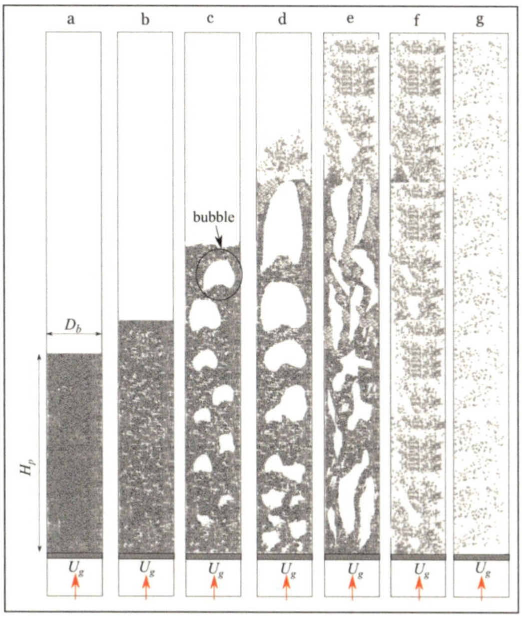

Figure 4.
Schematic illustration of the fluidization regimes.

And, theoretically, as the gas velocity increases, the pressure drop per bed weight remains constant. The shift in the bed height (bed expansion) due to the minimum fluidization condition is demonstrated in **Figure 4b**. As U_g continues to rise, the bed gains momentum, and motion starts with the bubbles emerging, as seen in **Figure 4c**. Maintaining a homogeneous bubble distribution, bubble formation, and bubble frequency is critical for better mixing, which leads to higher heat transfer rates and, consequently, faster reaction rates. As the gas velocity increases, the bubbles condense into larger bubble formations known as "slugs" with a diameter similar to the bed diameter (**Figure 4d**). Bubble formations break up as U_g increases, resulting in rapid particle mixing (**Figure 4e**). A higher increase in the gas velocity causes a transition to a fast fluidization regime (**Figure 4g**). A further increase in the gas velocity can transport the particles outside of the bed, as seen in **Figure 4g**. This regime is called pneumatic transport.

1.4 Geldart's particle classification

The multiphase flow fluidization process is influenced by particle diameter, sphericity, and density. Geldart [21] classified four particle groups according to their fluidization behavior by measuring the variations in gas and solid phase densities with the mean particle diameter. Geldart group particles are classified as follows:

Group A: Particles in Group A, such as cracking catalysts, have a tiny diameter (20 to 100 μm) and/or low density. After the minimum fluidization condition, dense phase expansion is visible. As a result, Group A particles require higher gas velocities for bubble formation compared to Group B particles.

Group B: Group B particles have mean diameters and densities ranging from 40 to 500 μm and 1.4 to 4 g/cm^3, respectively. A good illustration of this category of particles is sand. Bubbles are visible shortly after the minimum fluidization velocity. This group of particles shows the best fluidization characteristics.

Group C: Group C refers to particles having a diameter (10 to 80 μm) and a high degree of cohesion. Due to the increased interparticle forces caused by their high cohesive nature, they mix and fluidize poorly. Bubbles can be seen shortly after the lowest fluidization velocity, with a slight bed expansion.

Group D: Group D particles often have high particle densities and have diameters greater than 600 μm. Hence, they require higher gas flow rates to fluidize. Compared to Group B particles, they show poorer fluidization behavior.

2. Experimental setup and methodology

2.1 Cold flow and BFBG test rigs

The experimental setup consists of the cold flow (**Figure 5a**) and BFBG (**Figure 5b**) test stands. The cold flow test rig, made of transparent acrylic, allows the visualization of the fluidization behavior and the measurement of the pressure drop at ambient conditions. The main parts of the cold flow test rig are a plenum, a distributor plate, the bed section, and the freeboard (**Table 2**). A stainless steel distributor plate with a 10-μm pore size and 39% total porosity was used in the test rig. A mass flow controller was used to control and regulate the flow rate of the fluidizing gas (nitrogen, air). Pressure was measured at nine different points aligned vertically with

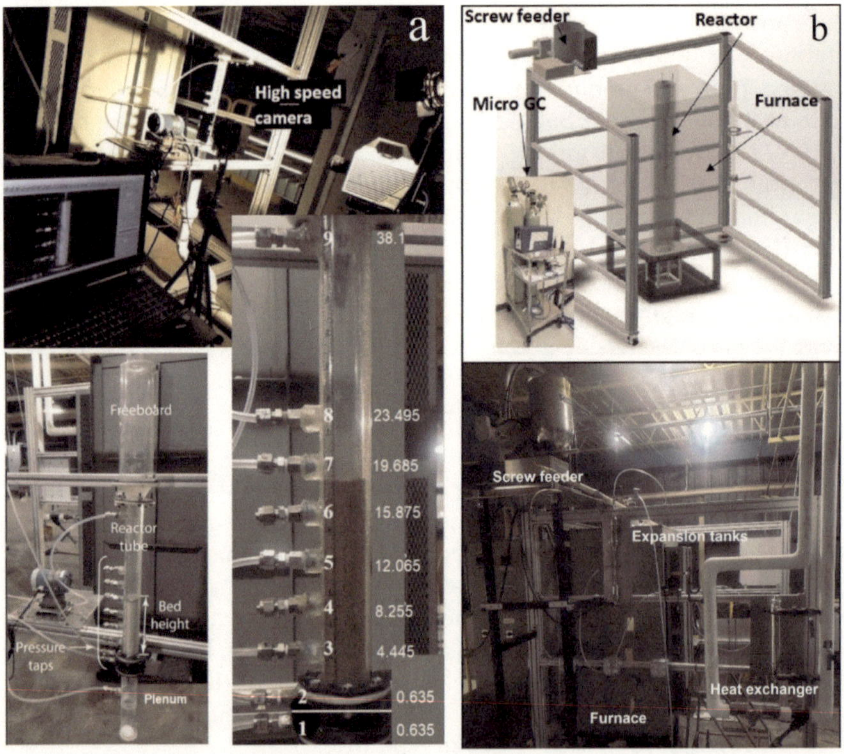

Figure 5.
a) Cold flow and b) BFBG experimental setups.

Material	Polymethyl-methacrylate
Reactor internal diameter (cm)	3.81
Reactor tube height (cm)	40.64
Transition cone height (cm)	2.54
Freeboard internal diameter (cm)	7.62
Freeboard height (cm)	25.4

Table 2.
Cold flow rig characteristics.

3.81 cm increments along the bed height, starting just below the distributor plate (measurement point 1) up to the measurement point 9 just below the transition cone between the reactor bed and the freeboard. Pressure signals were recorded at a 10 Hz sampling rate using a data acquisition system. Later, the obtained data was analyzed with Python-based software. **Figure 5a** depicts the alignment of the pressure taps. Fluidization behavior was studied by using the images captured by a high-speed camera for each test case. Images were processed with an open-source image processing tool called Python-Scikit to improve their qualities and the contrast between the inert and feedstock materials to better visualize the mixing condition. Cold flow experiments were conducted for the total mixture masses of 100, 200, and 300 g with the feedstock (coal) on top (segregated state) with a weight ratio of 4% to simulate the actual BFBG hydrodynamical behavior at the elevated temperatures. For each test case, measurements were taken after waiting at least another 30 seconds to stabilize the test case and avoid transition data. Tests were repeated three times, and the results are shared in the Cold Flow Analysis section.

The BFBG test stand consists of the BFBG reactor, a top-load furnace, a screw feeder, and a micro gas chromatograph to analyze the gas composition of the synthetic gas acquired from the gasification tests. A BFBG reactor was installed inside the furnace, which can reach temperatures of 1500°C. The temperatures inside the reactor, just above the inert material, and on the wall were measured with K-type thermocouples. The maximum temperature measured during the experiments was around 820°C, which is adequate to provide the required heat transfer rates for a successful gasification reaction. The BFBG reactor was made of inconel steel with similar dimensions as the cold flow test rig. The BFBG test stand also measured pressure drop to investigate the effect of elevated temperatures on bed pressure drop. Tests were conducted at the same time interval for the 200 and 300-g unary sand mixtures. The study conducted by Sivri [3] contains a detailed description and information about the design of the test stands.

2.2 Material analysis and preparation

In this study, the cold flow and actual BFBG tests used coal as the feedstock and sand or glass beads as the bed material, respectively. It was Pittsburgh coal seam number eight that was used as feedstock. Glass beads from Ballotini and commercial-grade fine silica sand from Quikrete brands were used as bed materials. The results of moisture, volatile, ash, and elemental analyses of the biomass and coal are displayed in **Table 3**. Further, the size and sphericity analyses with their distributions were

	Carbon (C), (%)	Hydrogen (H), (%)	Oxygen (O), (%)	Sulfur (S), (%)	Moisture (%)	Ash (C), (%)
Coal, Pittsburgh #8	73.62	4.38	7.83	2.59	3.69	7.89

Table 3.
Elemental and proximate analysis (by mass) of bituminous coal.

Material	Average of sphericity	Sauter mean diameter (μm)	Bulk density (g/cm^3)	Skeletal density (g/cm^3)	Geldart's group
Coal	0.85	362	0.7	1.36	B
Sand	0.86	324	1.43	2.64	B
Glass beads	0.93	271	1.46	2.48	B
Coal (wt%4) and Sand	0.86	325	1.56	2.59	B
Coal (wt%4) and Glass beads	0.93	282	1.61	2.44	B

Table 4.
Material size, density, sphericity, and Geldart's group analysis.

conducted with a dynamic image analysis method (Sympatec GbmH, Model QICPIC) and skeletal densities were analyzed by a gas pycnometer (AccuPyc, Model 1330 Helium Pycnometer). The results of the size, sphericity, and density analyses for the feedstock and inert materials are summarized in **Table 4**. Glass beads showed the highest mean sphericity of 0.93 compared to 0.85 for coal, and 0.86 for sand, respectively. Sand and glass beads have a narrower sphericity (90% between 0.85 and 0.95) and size (90% between 235 and 347 μm) distribution compared to coal as well. Coal has a bigger mean diameter of 362 μm compared to glass beads (271 μm) and sand (324 μm). Hence, the coal and glass beads mixture has better packing, which enhances the fluidization characteristics (**Figure 6**).

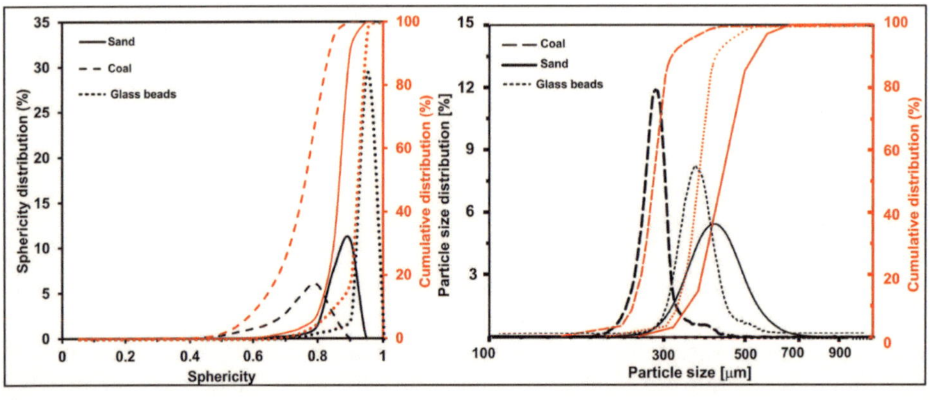

Figure 6.
Particle size and sphericity distributions.

3. Cold flow analyses

BFBG deep-bed applications have a higher gasification efficiency due to improved heat transfer rates and a longer gas residence time. However, top-fed deep-bed application of the BFBG is quite complicated, not only because of the requirement to deliver the feedstock particles homogeneously on top of the bed, but also because of the requirement for an optimum homogeneous binary mixture to obtain the most efficient gasification process. Hence, the observation of the fluidization process in cold flow conditions is required to better understand the intricate hydrodynamics of coal-top-fed deep-bed $(H_p/D_b \gtrsim 2)$ binary mixtures.

3.1 Mixing and fluidization behavior of coal with sand or glass beads mixtures

The mixing and fluidization behavior of binary mixes of coal with two separate inert components (glass beads and silica sand) was analyzed in this section. The total masses of the binary mixtures investigated were 100, 200, and 300 g, respectively, including coal, which made up 4% of the total mass and was spread on top of the bed material. Bed pressure drop, ΔP_b, and minimum fludization velocity, U_{mf}, were analyzed as a function of superficial fluidizing gas velocity with a 0.0146 m/s (1 SLM) increments. The bed fluidization and mixing behavior were also observed with the high-speed camera at each flow rate after attaining fluidization. In **Figures 7–12**, each column picture under the pressure drop curve represents the bed behavior at the corresponding fluidizing gas velocity. With the use of these images, the ideal fluidizing-gas velocity interval was estimated to obtain an almost homogeneous (quasi-homogeneous) binary mixture with reliable fluidization behavior for the BFBG operation.

Coal and glass beads binary mixtures have higher bulk density $(1.61 \text{ g/cm}^3 > 1.56 \text{ g/cm}^3)$ and higher average sphericity $(0.93 > 0.86)$ (**Table 4**) compared to coal with sand mixtures. Hence, better fluidization and mixing behaviors were expected for the coal and glass beads mixtures. Later on, Case I and Case II represent the coal and glass beads mixture and the coal and sand mixture, respectively. The results obtained during the investigation of the mixing and fluidization behavior of the binary mixtures are shared in this section.

The bed pressure drop and fluidization behaviors of the binary mixtures of Case I and Case II for the total mixture mass of 100 g with increasing fluidizing gas velocity are shown in **Figures 7** and **8**, respectively. The initial static bed aspect ratio (H_p/D_b) measured as ≈1.83 for Case I, and slightly higher ≈2 for Case II due to the lower average sphericity and bulk density. In Case I, there was a smooth transition from the minimum fluidization to complete fluidization after the pressure curve climbed linearly up to the initiation of the fluidization. However, the transition was not smooth in Case II. The abrupt drop in bed pressure drop was followed by an increase in U_g, and ΔP_b rose until all particles fluidized as a result of Case II's greater particle size distribution and channeling. Tiny bubble formations due to channeling before complete fluidization were visible in Case II at the speeds between 0.117 and 0.146 m/s (**Figure 8b** columns 5–7). Besides, wider particle size distribution and less sphericity encoupled with the humidity effects of Case II caused to reach minimum fluidization velocity at a higher speed of 0.06 m/s compared to Case I with a $U_{mf} \approx 0.05$ m/s. As seen in **Figure 7b**, except for a tiny layer of coal on top of the glass beads in Case I, the bed was almost in a well-mixed state for the fluidization gas velocities of 0.088 and

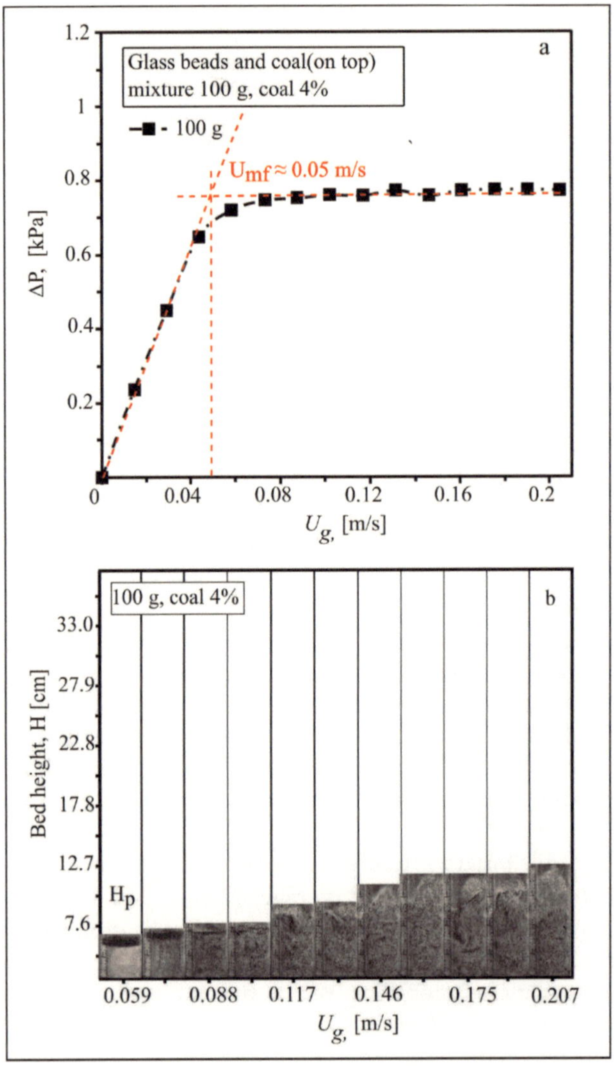

Figure 7.
Bed pressure drop and fluidization behaviors with increasing superficial gas velocity for the mixtures of coal and glass beads with the total mixture mass of 100 g.

0.103 m/s (**Figure 7b** third and fourth columns, respectively). A further increase in the flow rate resulted in smooth bubble formations and eventually a well-mixed state at around $U_g = 0.12$ m/s. However, Case II required a higher U_g of 0.16 m/s to achieve a well-mixed state (**Figure 8b** eighth column) due to the same reasons mentioned in the U_{mf} comparison for both cases. In Case II, mixing happened just after reaching the complete fluidization velocity (U_{cf}) of ≈0.14 m/s because of the sudden breakdown of the interparticle and cohesive forces. At higher gas velocities after reaching the complete fluidization, flat slug formations were not observed due to the low bed aspect ratios, which were around two for both cases.

Cases I and II were also tested for the total mixture mass of 200 g with the coal making up 4% of the total mass. **Figures 9** and **10** demonstrate the fluidization

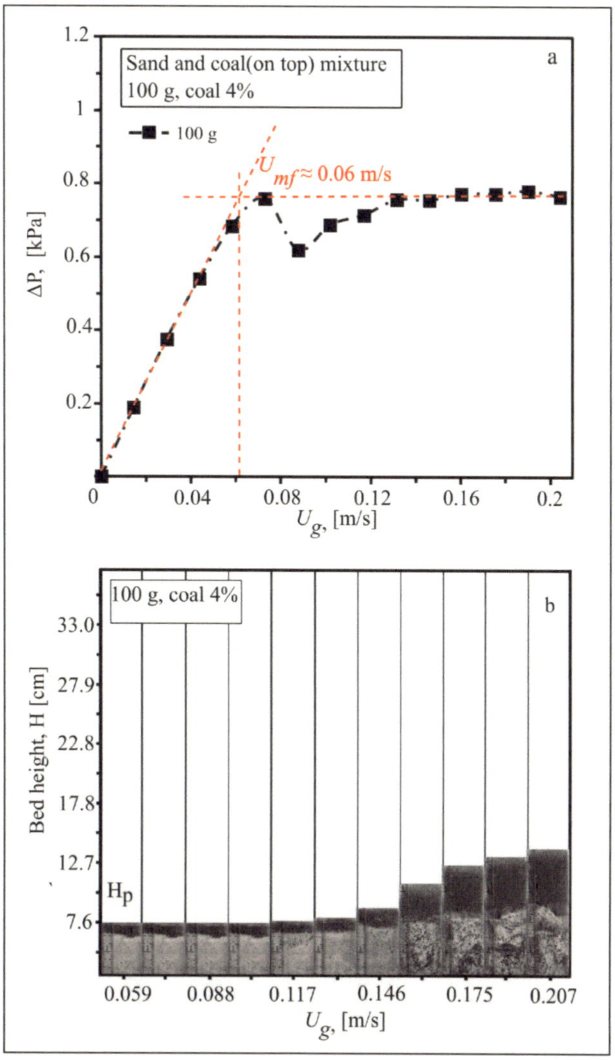

Figure 8.
Bed pressure drop and fluidization behaviors with increasing superficial gas velocity for the mixtures of coal and sand with the total mixture mass of 100 g.

behavior and pressure drop changes with increasing gas velocity for Cases I and II, respectively. Static bed aspect ratios (H_p/D_b) were 3.5 and 3.66 for Cases I and II, respectively. In both Cases, the bed pressure drop showed a linear increase during the fixed-bed state. As in the previous test, the transition from fixed bed state to fluidization was smoother for Case I compared to Case II due to the reasons mentioned earlier. Furthermore, in Case I, mixing began earlier at $U_g{\approx}0.074$ m/s, just after reaching the complete fluidization, with the penetration of glass beads into the coal layer. $U_g{\approx}0.088$ m/s achieves a nearly well-mixed state, except for the tiny coal layer on top of the bed. But, in Case II, complete fluidization was achieved relatively late at a gas velocity of $U_g{\approx}0.09$ m/s, and complete mixing could be achieved at the gas velocity of 0.103 m/s with a narrow coal layer on top. Another factor, except the sand characteristics, that contributed to the delay in reaching the complete fluidization and

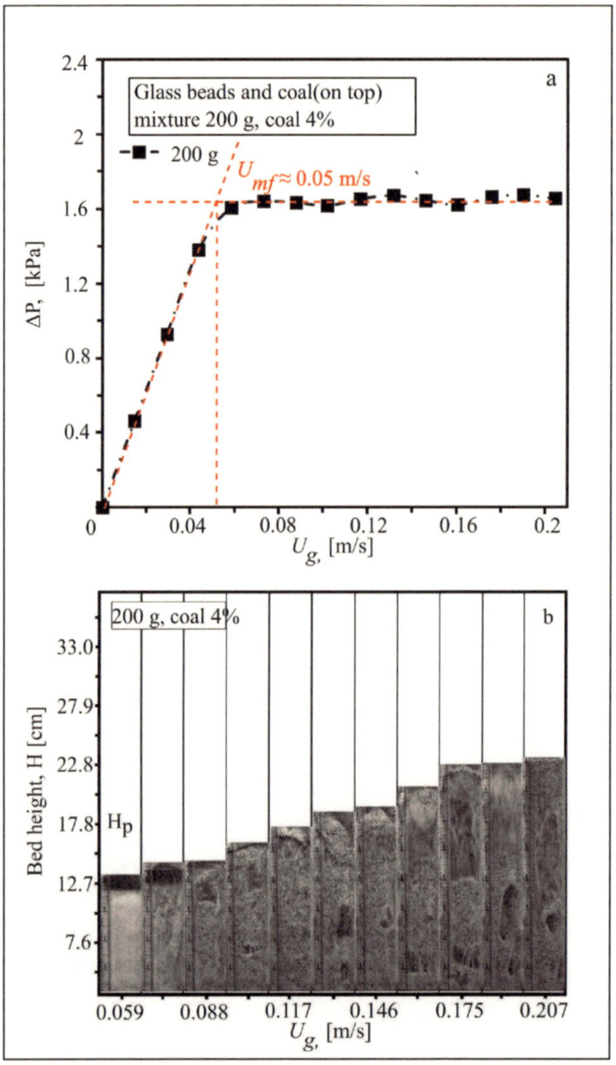

Figure 9.
Bed pressure drop and fluidization behaviors with increasing superficial gas velocity for the mixtures of coal and glass beads with the total mixture mass of 200 g.

well-mixed state was the higher relative humidity of the fluidizing gas (air), which strengthened the interparticle forces and cohesiveness. In both cases, U_{mf} was measured graphically around 0.05 m/s. Further increase in U_g caused slug formations, which can be seen in both Cases. Despite the slug formations, the well-mixed states achieved in both Cases (**Figure 9b** and **10b**, columns 4–11), supported by the same images, demonstrate almost a homogeneous distribution of the coal particles inside the bed material. Some of the coal dust particles were stuck on the bed wall in Case II due to the higher ambient humidity.

Later on, experiments were conducted to test the fluidization behaviors for a deeper bed with bed aspect ratios of 5.16 and 5.66 and a total mass of 300 g, with the coal making up 4% of the total mass for Cases I and II, respectively. Fluidization behaviors and pressure drop attitude with increasing gas velocity for Cases I and II

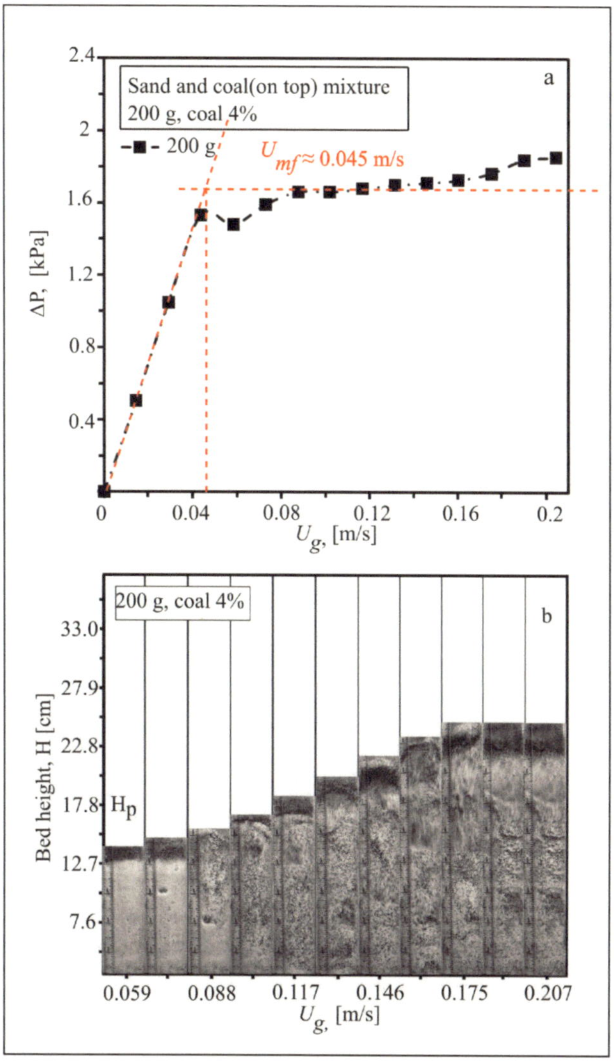

Figure 10.
Bed pressure drop and fluidization behaviors with increasing superficial gas velocity for the mixtures of coal and sand with the total mixture mass of 200 g.

with the total mass of 300 g are shown in **Figures 11** and **12**, respectively. Similarly, bed pressure drop increased linearly during the fixed-bed state for both Cases as in the previous test conducted with the total masses of 100 and 200 g, respectively. Despite the higher bed aspect ratio compared to the previous tests, Case I still showed a smooth transition from static state to fluidization as seen in **Figure 11**. On the contrary, transition was relatively stiff for Case II. Due to the stronger interparticle forces and cohesiveness in Case II peak pressure drop value obtained was higher (≈3.3 kPa) compared to Case I (≈2.4 kPa). After reaching the peak pressure drop in Case II, there was an abrupt fall to ≈2.4 kPa. Case I and II reached the complete fluidization states at the approximate gas velocities of ≈0.07 m/s and ≈0.075, respectively. For both Cases after reaching the complete fluidization, coal particles started mixing with the sand particles at the contact of the segregated layers. Complete mixing is achieved at an approximate gas velocity of 0.09 m/s for Case I, and 0.1 m/s for Case II. As in the

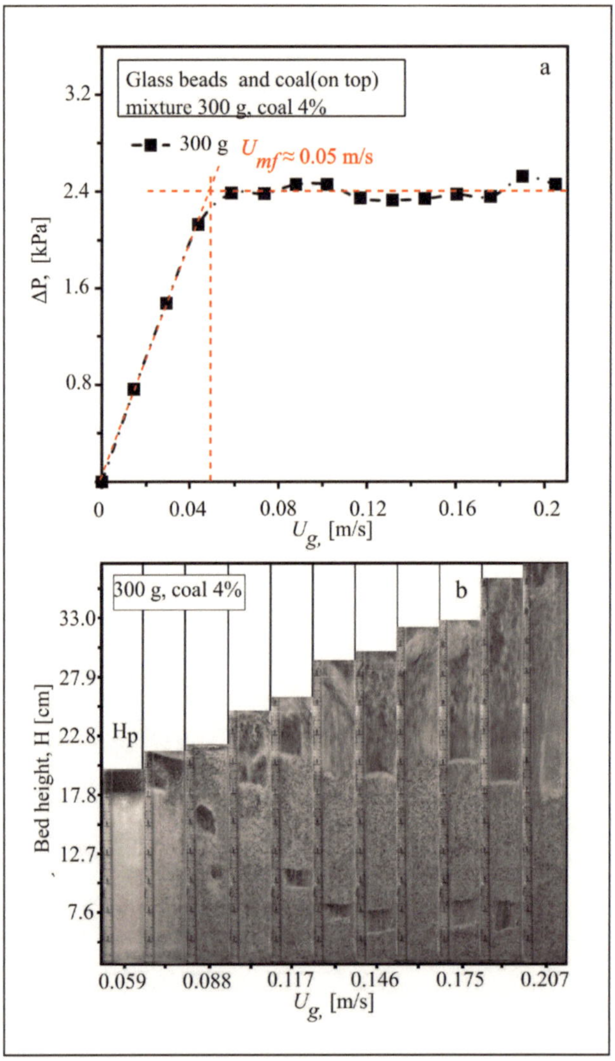

Figure 11.
Bed pressure drop and fluidization behaviors with increasing superficial gas velocity for the mixtures of coal and glass beads with the total mixture mass of 300 g.

previous experiments conducted for 100 and 200 g total masses, a well-mixed state was achieved at a higher gas velocity in Case II compared to Case I for 300 g due to the same reasons mentioned before. Slug formations were visible for both Cases after the gas velocity of ≈0.1 m/s (**Figure 11b** and **12b**, columns 4–11). For both Cases, the bed could preserve its well-mixed state at higher gas velocities despite the narrow coal layer on top of the bed. Elutriation was observed at a gas velocity of 0.175 m/s and higher for Case II, and 0.207 m/s for Case I.

3.2 Summary and conclusions

The fluidization and mixing behaviors were investigated for Cases I (coal and glass beads) and II (coal and sand) mixtures with different total masses of 100, 200, and

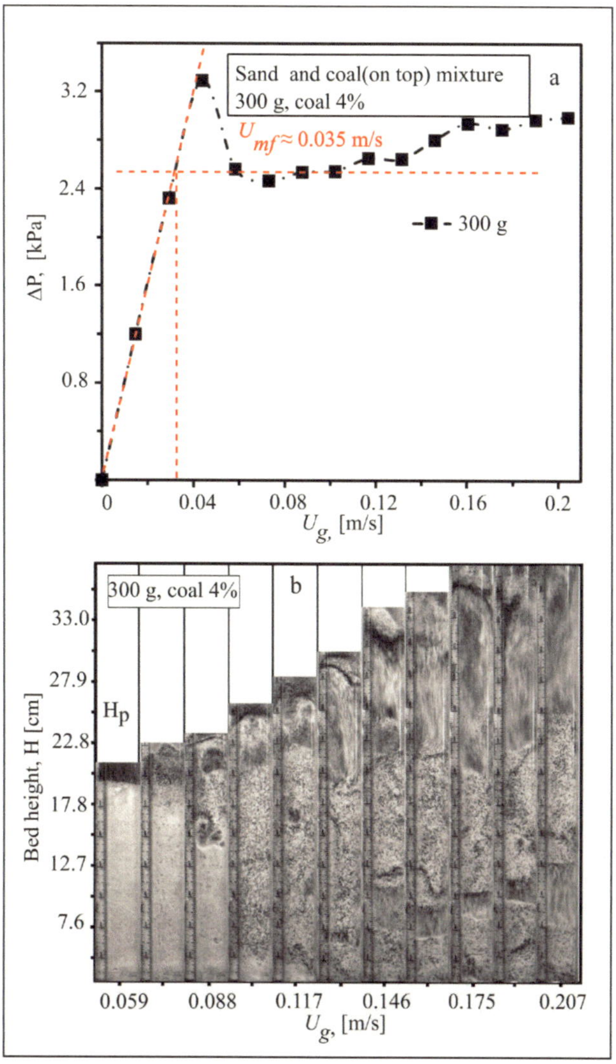

Figure 12.
Bed pressure drop and fluidization behaviors with increasing superficial gas velocity for the mixtures of coal and sand with the total mixture mass of 300 g.

300 g, with coal making up 4% of the total mass. Initially, the mixtures were in a segregated state in which coal was placed on top. The fluidization and mixing behaviors of the mixtures were observed as the gas velocity increased. The main conclusions drawn from this study were:

- Coal and glass beads mixtures showed better fluidization and mixing characteristics compared to coal and sand mixtures at each total mass tested due to their higher bulk density and average sphericity, which led to better packaging.

- Higher average sphericity and bulk density led coal and glass bead mixtures to reach complete fluidization and well-mixed conditions at lower gas velocities.

- Strong interparticle interactions between sand particles may result in higher peaks of bed pressure drop and cohesiveness, particularly when the fluidizing gas is air with a higher relative humidity than usual.
- Slug formations, which decrease mixing and fluidization quality, were more likely to be formed with an increasing bed aspect ratio greater than two.
- With increased gas velocity, both mixtures in Cases I and II could achieve and maintain a well-mixed state at bed aspect ratios greater than two.

4. Fluidization and gasification analyses at elevated temperatures

The effect of temperature on fluidization characteristics such as bed pressure drop and minimum fluidization velocity was investigated, and the results are shared in this section for sand material. Besides, gasification test results for the binary mixture of coal and sand are shared.

4.1 Effect of temperature on the bed pressure drop and minimum fluidization velocity

The bed pressure drop versus gas velocity was studied for unary sand mixtures of 200 and 300 g at elevated temperatures ranging from 200°C to 805°C (**Figure 13**). The bed aspect ratios measured were 3.3 and 4.83 for the total masses of 200 and 300 g, respectively, in the cold flow test rig, with similar dimensions to the BFBG.

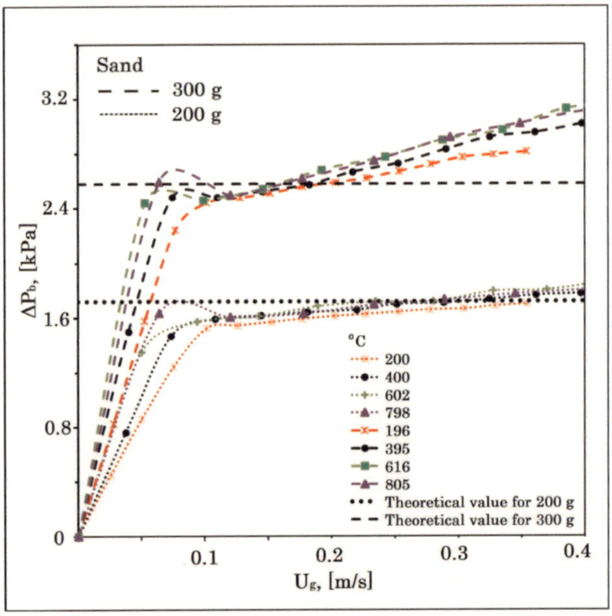

Figure 13.
Bed pressure drop versus fluidization gas velocity at different temperatures. G.

Bed pressure drop increased linearly at all temperatures for both total bed masses of 200 and 300 g, respectively. A linear increase in the pressure drop was followed by a smooth transition to complete fluidization in all cases. However, for the mass of 300 g, small pressure drop peak formations were visible due to higher wall effects. After complete fluidization, the pressure drop continued to increase with the increasing gas velocity due to three main factors: wall effects, increased bed voidage [22], and stronger interparticle forces particularly for this narrow-sized (3.81 cm diameter) deep-bed reactor. Similar results of increasing bed pressure drop due to wall effects were also reported by Srivastava and Sunderasan [23] and Olatunde et al. [24]. While the minimum fluidization condition's bed pressure drop increased with temperature, U_{mf} decreased for both total masses of 200 and 300 g (from 0.086 to 0.063 m/s for 200 g, and from 0.096 to 0.062 m/s for 300 g). The decrease in the U_{mf} can be attributed to the higher interparticle forces [22], and lower *Re* numbers at higher gas temperatures, which lead to higher gas viscosity and increase the drag coefficient.

4.2 Gasification analysis

The findings of the gasification of coal with 10% steam and air in the BFBG are presented in this subsection. The elemental composition of coal (**Table 3**) reveals that it has a low oxygen content, demanding the use of an outside oxygen source to create the appropriate syngas composition. As a fluidizing agent, a mixture of air and steam was used. And sand was preferred due to its lower thermal conductivity for a better thermal management of the reactor. When 10% steam was added to the coal feed, an average ratio of $H_2/CO = 3.23$ was observed ignoring the nitrogen content. Gasification tests were performed in a slightly fast fluidization regime to assure sudden mixing and enhanced particle-particle and particle-gas interaction, which generates better heat transfer between the bed material (sand) and the feedstock (coal) particles and improves reaction kinetics. The complex gasification reaction kinetics interact strongly with fluidization hydrodynamics, particularly bubble formations; shape, frequency, and so on. As a result, fluidization hydrodynamics should be studied in greater depth in order to better understand the complicated process of gasification and its interaction with fluidization hydrodynamics. Initially, the temperature of the bed medium was elevated to the desired temperatures (around 800°C) to achieve an effective gasification process. Temperature was measured on the reactor wall, at the level of the bed material, and inside the bed, just above the bed material, to assure an accurate temperature measurement to start the gasification process. Later, for each supply, 2 g of coal were fed by the supply line using pressurized nitrogen. Coal flow rate was maintained at the rate of 2 g/min. The average syngas composition at the initial stage of the pyrolysis, that liberates hydrogen component in coal first before proceeding the carbon gasifying status, and its low heating value obtained during the tests are shared in **Table 5**.

H_2, (%)	CO, (%)	CO_2, (%)	CH_4, (%)	LHV, ($MJNm^{-3}$)
50	16	17	15	11.72

Table 5.
Syngas composition and low heating value.

4.3 Summary and conclusions

The effect of the temperature on the BFBG fluidization hydrodynamics was analyzed and the initial gasification test results were shared in Chapter 4. The main conclusions are:

- The minimum fluidization velocity decreases with the increasing temperature. It can be related to the stronger interparticle forces and lower wall effects with the increasing temperature.

- The gasification results show that syngas can be generated successfully by using the top-fed feeding system. And the product gas can be used as primary or dual fuel in internal combustion engines and solid oxide fuel cell applications.

Acknowledgements

This work was supported by US Department of Energy's Fossil Energy Advanced Gasification Program. The Research was executed through NETL Research and Innovation Center's Advanced Gasification effort within the Advanced Reaction Systems FWP under the RSS contract 89243318CFE000003. The author gratefully acknowledge WVU's CIGRU, MAE, and CBE departments for their assistance with comprehensive mechanical, electronic, and data-acquisition hardware/software systems, respectively. The research was conducted at the WVU's Advanced Combustion Laboratory in Morgantown, WV.

Author details

Ali Can Sivri
Department of Mechanical and Aerospace Engineering, West Virginia University, Morgantown, USA

*Address all correspondence to: acs0031@mix.wvu.edu

References

[1] Bates RP, Dölle K. Syngas use in internal combustion engines-a review. Advances in Research. 2017;**10**(1):1-8. DOI: 10.9734/AIR/2017/32896

[2] Abdullah M, Husain Z, Pong SY. Analysis of cold flow fluidization test results for various biomass fuels. Biomass and Bioenergy. 2003;**24**(6): 487-494. DOI: 10.1016/S0961-9534(02) 00150-2

[3] Sivri AC. A Study on the Hydrodynamics of a Bench-Scale Top-Fed Bubbling Fluidized Bed Gasifier Using Biomass and Coal as Feedstocks. Graduate Theses, Dissertations, and Problem Reports 7907; 2020. Available from: https://researchrepository.wvu.edu/etd/7907

[4] Zhang Y, Zhong W, Jin B. Experimental and theoretical study on fluidization of stalk-shaped biomass particle in a fluidized bed. International Journal of Chemical Reactor Engineering. 2011;**9**(1). DOI: 10.2202/1542-6580.2691

[5] Clarke K, Pugsley T, Hill G. Fluidization of moist sawdust in binary particle systems in a gas–solid fluidized bed. Chemical Engineering Science. 2005;**60**(24):6909-6918. DOI: 10.1016/j.ces.2005.06.004

[6] Li J, Kwauk M, Li H. Fluidization of nonspherical particles: A review. Powder Technology. 2006;**162**(1):1-19. DOI: 10.1016/j.powtec.2005.12.006

[7] Amiri Z, Movahedirad S. Bubble-induced particle mixing in a 2-D gas-solid fluidized bed with different bed aspect ratios: A CFD-DPM study. Powder Technology. 2017;**320**:637-645. DOI: 0.1016/j.powtec.2017.07.097

[8] Fu Z, Zhu J, Barghi S, Zhao Y, Luo Z, Duan C. Mixing and segregation behavior in an air dense medium fluidized bed with binary mixtures for dry coal beneficiation. Powder Technology. 2020;**371**:161-169. DOI: 10.1016/j.powtec.2020.05.094

[9] Gao Z, Zhang J, Wang J. Effect of particle density on the hydrodynamics of gas-solid fluidized beds with Geldart group B particles. Powder Technology. 2009;**192**(1):32-40. DOI: 10.1016/j.powtec.2009.01.016

[10] Abdullah MS, Karim MIA, Aziz ARA. Effect of mixture bulk density and bed voidage on minimum fluidization velocity of various materials. Powder technology. 2013;**245**:177-184. DOI: 10.1016/j.powtec.2013.05.013

[11] Escudero D, Heindel TJ. Bed height and material density effects on fluidized bed hydrodynamics. Chemical Engineering Science. 2011;**66**(16): 3648-3655. DOI: 10.1016/j.ces.2011.05.027

[12] Sánchez-Delgado S, Almendros-Ibáñez JA, Garcá-Hernando N, Santana D. On the minimum fluidization velocity in 2D fluidized beds. Powder Technology. 2011;**207**(1-3):145-153. DOI: 10.1016/j.powtec.2010.12.032

[13] Anantharaman A, Cocco RA, Chew JW. Evaluation of correlations for minimum fluidization velocity (U_{mf}) in gas-solid fluidization. Powder Technology. 2018;**323**:454-485. DOI: 10.1016/j.powtec.2017.12.071

[14] Shabanian J, Chaouki J. Effects of temperature, pressure, and interparticle forces on the hydrodynamics of a gas-solid fluidized bed. Chemical Engineering Journal. 2017;**313**:580-590. DOI: 10.1016/j.cej.2016.12.060

[15] Milacic E, Nunez Manzano M, Madanikashani S, Heynderickx GJ, van Geem KM, van de Greef A, et al. Experimental study on the temperature distribution in fluidised beds. Chemical Engineering Science. 2022;**248**:117062. DOI: 10.1016/j.ces.2021.117062

[16] Lim MT, Alimuddin Z. Bubbling fluidized bed biomass gasification-performance process findings and energy analysis. Renewable Energy. 2008;**33**(10):2339-2343. DOI: 10.1016/j.renene.2008.01.014

[17] Narvaez I, Orio A, Aznar MP, Corella J. Biomass gasification with air in an atmospheric bubbling fluidized bed. Effect of six operational variables on the quality of the produced raw gas. Industrial & Engineering Chemistry Research. 1996;**35**(7):2110-2120. DOI: 10.1021/ie9507540

[18] Cui H, Grace JR. Fluidization of biomass particles: A review of experimental multiphase flow aspects. Chemical Engineering Science. 2007;**62**(1–2):45-55. DOI: 10.1016/j.ces.2006.08.006

[19] Anantharaman A, Cocco RA, Chew JW. Evaluation of correlations for minimum fluidization velocity (U_{mf}) in gas-solid fluidization. Powder Technology. 2018;**323**:454-485. DOI: 10.1016/j.powtec.2017.10.016

[20] Ergun S. Fluid flow through packed columns. Chemical Engineering Progress. 1952;**48**(2):89-94

[21] Geldart D. Types of gas fluidization. Powder Technology. 1973;7(5):285-292. DOI: 10.1016/0032-5910(73)80037-3

[22] Formisani B, Girimonte R, Mancuso L. Analysis of the fluidization process of particle beds at high temperature. Chemical Engineering Science. 1998;**53**(5):951-961. DOI: 10.1016/S0009-2509(97)00370-9

[23] Srivastava A, Sundaresan S. Role of wall friction in fluidization and standpipe flow. Powder Technology. 2002;**124**(1–2):45-54. DOI: 10.1016/S0032-5910(01)00471-5

[24] Olatunde G, Fasina O, McDonald T, Adhikari S, Duke S. Moisture effect on fluidization behavior of loblolly pine wood grinds. Biomass Conversion and Biorefinery. 2017;7(2):207-220. DOI: 10.1007/s13399-016-0223-9

Chapter 6

Transformation of Waste Coal Fly Ash into Zeolites for Environmental Applications

Henilkumar M. Lankapati, Kalpana C. Maheria and Ajay K. Dalai

Abstract

The generation of a large quantity of waste coal fly ash (CFA) *via* coal combustion process during power generation is of major concern as disposal of such huge quantity of fly ash causes serious threats to the environment. There is an exigent need to find out the proper solution for its disposal/utilization to reduce its harmful effects. The composition of waste coal fly ash mostly consists of silica and alumina. Hence, the researchers are tempted to utilize waste coal fly ash as a starting ingredient to make value-added materials like zeolites. It is anticipated that such research efforts will act as a valuable aid to reduce the disposal cost of fly ash and ultimately reduce harmful effects of fly ash to the environment. In this review, various synthesis methods to synthesize different types of zeolites from CFA, such as Zeolite-A, Zeolite-X and Zeolite-P, have been summarized and their potential for various applications such as sorption and catalysis has been explored.

Keywords: fly ash, waste utilization, zeolites, catalysis, sorption

1. Introduction

Environmental issues are the most burning issues in the world nowadays. The leading industrialist, scientists, researchers, and environmentalist around the world strive hard for the finding of the proper solution to mitigate or reduce the impact of various environmental problems. The rising issues and researches concerning around the world are (i) to reduce CO_2 levels by utilizing renewable sources as fuels and for the energy production, (ii) storage and disposal of the waste coal fly ash generated from the thermal power plants, (iii) to identify the various contaminants such as toxic/heavy metals, pesticides, petroleum hydrocarbons, which deteriorate the air, soil, and water environments, (iv) to reduce the generation of various pollutants by adopting green chemistry principles, and (v) to find out the best treatment technologies for the removal of contaminants and recovery of the precious metals or products from the waste materials or to convert the waste materials to value-added materials.

Nowadays, environmental researchers and scientists focus on the utilization of waste for the treatment of waste, that is, converting waste materials into value-added materials such as adsorbents and catalysts and utilize them for the effective treatment of the waste. The waste coal fly ash (CFA) is generated as a byproduct during the electricity generation in thermal power plants. The CFA is considered as the harmful materials, and it is generated in a vast quantity that their disposal is so crucial and required large acre of land. The composition of waste CFA mostly consists of silica and alumina along with trace amount of other rare earth elements, which are precious. Hence, the researchers are tempted to utilize waste coal fly ash as a source of precious elements and make value-added materials such as zeolites from it. Synthesis of different types of zeolites from CFA, such as zeolite X, P, and A, is reported in the literature [1, 2].

Few studies have been undertaken for the applications of CFA-derived materials as the environmental remediation, such as, removal of phosphates and nutrients from the water [3]. Deng et al. have reported the method of converting CFA into materials like adsorbent using microwave-assisted alkali modification and utilized to adsorb Cr^{6+} metal ions from the water [4]. Appiah-Hagan et al. have modified waste CFA using freezing and thawing method and studied the adsorption efficiency of modified fly ash for the sorption of Pb^{2+}, Co^{2+}, Ni^{2+}, Cu^{2+} Cr^{3+}, and Cd^{2+} [5]. Similarly, lime-activated fly ash is utilized for the sequestration of heavy metal ions, such as Zn^{2+}, Pb^{2+}, and As^{5+} from the aqueous solution [6]. Chinh et al. have explored the potential of fly ash modified with (3-mercaptopropyl) triethoxysilane for the scavenging of Hg^{2+} ion [7]. Lathiya et al. have converted waste CFA into sulfated fly ash and applied it as a solid acid catalyst for the biodiesel production from maize acid oil [8]. These studies reveal that fly ash can be utilized efficiently for its conversion into valuable materials and may be used further for metal extraction from aqueous solution.

2. Fly ash generation and utilization

2.1 Fly ash generation

Thermal power plants are the main sources of the waste CFA generation as a byproduct of electricity generation. According to the bp (British Petroleum) statistical review of world energy 2022 (71st edition) published by British Petroleum, the generation of electricity is increased by 6.2% during the year 2021. Along the total generation of electricity, coal remains the dominant as a fuel with 36% of its share followed by natural gas (23%), hydroelectricity (15%), renewable sources (13%), nuclear energy (10%), and others (1%) as shown in **Figure 1** [9].

The last 6-year scenario of the electricity generation by fuels as shown in **Figure 2**, reveals that the coal remains dominant (approx. 36%) over all other sources since last few years [9]. The coal utilization ultimately leads to the generation of huge quantity of waste CFA, and its disposal imposes threat to the environment due to its hazardous nature. Though energy generation modes are shifting toward renewable sources, still modes of thermal energy production dominate over the renewable sources as these technologies are under development stages. Hence, the dependency on the coal for energy production is still expected to dominate significantly in the upcoming years also. This will lead to generation of significant amount of fly ash, and there is an urgent need to develop the effective utilization modes of fly ash, which are economically and environmentally viable.

In India, around 76% of total energy demand is supplied by the coal-based thermal power plants. There are more than 202 coal-based thermal power plants in India, with

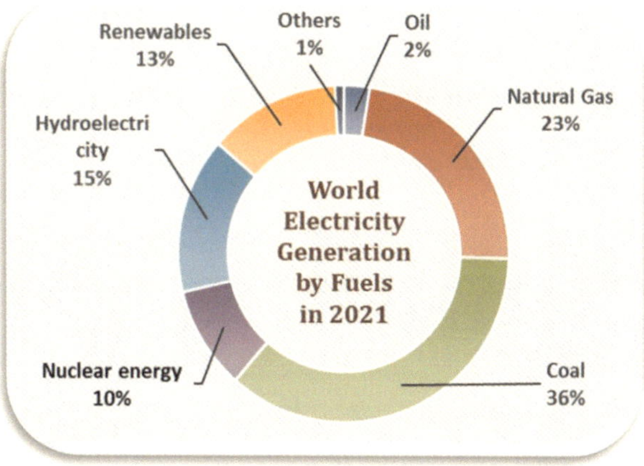

Figure 1.
World electricity generation by fuels during the year 2021 [9].

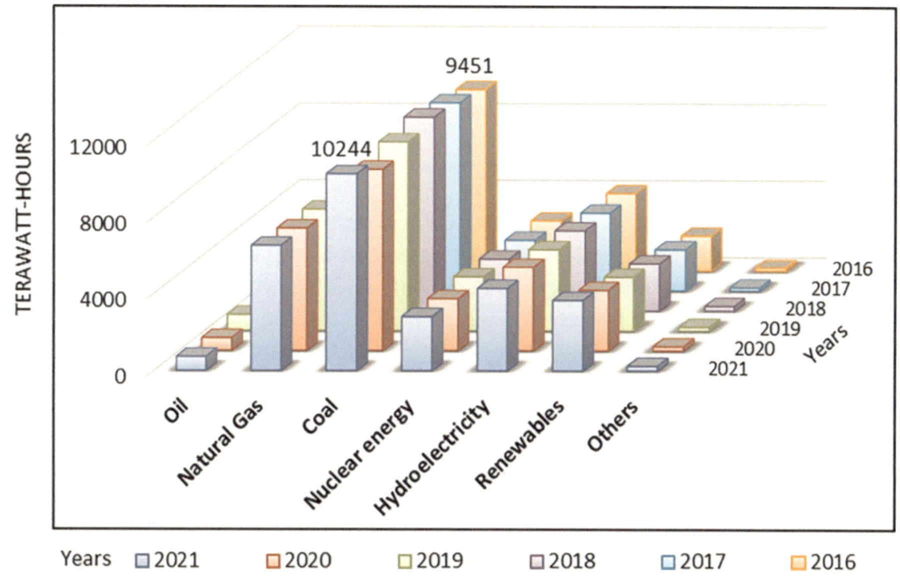

Figure 2.
Scenario of world electricity generation by fuels (terawatt-hours) from 2016 to 2021 [9].

installed capacity around 209,990 MW [10]. According to the CEA (Central Electricity Authority) report, August 2021, these power plants consume ~686 million tonnes of coal and produce ~233 million tonnes of waste CFA during the year 2020–2021. Annual fly ash generation from Indian coal power plants in last 10 years rose from 145.4 million tonnes during the year 2011–2012 to 232.6 million tonnes during the year 2020–2021, an increase of almost 60% [10].

According to the initiative implemented by Ministry of Environment, Forest and Climate Change (MoEFCC), Government of India, to utilize the 100% waste CFA, around 215 million tonnes of fly ash is utilized during the year 2020–2021, which is around 92.4% of total fly ash generation and 7.6% (~108.6 million tonnes) of fly ash

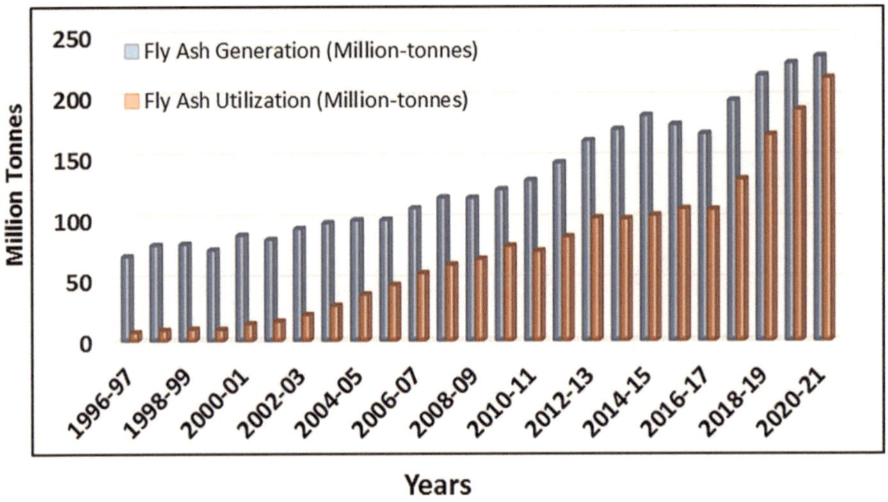

Figure 3.
Trends of CFA generation and utilization in India from 1996 to 2021 [10].

remains unutilized [10]. **Figure 3** shows the trends in fly ash generation and utilization in India from the year 1996 to 2021.

2.2 Legacy ash

The build-up of huge quantity of waste CFA by thermal power plants over decades or other terms, the fly ash stored in the ash ponds, or ash dykes by thermal power plants are considered as the legacy ash. From the CEA report, the available total quantity of legacy ash as on March 31, 2021 is 1738.19 million tonnes, which is dumped or stored at various ash dykes occupying large acre of land area surrounding the power plants [10].

According to the present practice in India, the requirement of land area for the disposal of coal fly ash by the year 2022 is estimated about 126,000 ha (approximately 0.6 ha per MW), if effective utilization will not be implemented. Even 92.4% fly ash is utilized, the land required for unutilized fly ash during the year 2020–2021 (∼108.6 million tonne) would be around ∼58,800 ha as per the latest reported data by CEA [10, 11].

2.3 Policy and notifications

MoEFCC, Government of India, have taken various initiatives for the protection of land to be polluted by the fly ash slurry, which is disposed into the ash dykes or ash ponds. In their effort to reduce the requirement of large acre of land, MoEFCC has issued first notification regarding the fly ash utilization on September 14, 1999, which was further amended time to time over the years, 2003, 2009, 2016, and 2021. The first notification includes the enforcement of CFA utilization for the manufacturing of bricks, tiles, blocks, and other construction-related materials, and manufacturer within 50 km radius of coal-based thermal power plant must mix at least 25% of CFA with soil by weight percentage [12].

In the first amendment, the radius of the availability of bricks/tiles manufacturers, who have to follow the notification was amended and expanded up to 100 km from

50 km. It further describes target for the waste fly ash percentages, which needs to be utilized by manufacturer as per the radius wise, up to year 2007 [13].

The amendment notification issued on November 3, 2009 provides the details of minimum quantity of fly ash, which needs to be utilized for building materials according to the category of fly ash based product. It also provides targets to the thermal power plants (those are already in operation before the date of this notification) for the utilization of fly ash in a phase manner to reach 100% utilization of fly ash, that is, starting from the date of the notification, target of 50, 60, 75, 90, and 100% for the first, second, third, fourth, and fifth, respectively, while for the newly constructed power plant, the target of achieving 50, 70, 90, and 100% fly ash utilization for the first, second, third, and fourth years, respectively, from the date of the commissioning of power plant [14].

Further, in the year 2016, the new amended notification issued in order to widen the scope of utilization and the vicinity of radius from 100 km to 300 km of the availability of manufacturers or builders needs to use the fly ash in building materials. It also stated that the cost of transportation of CFA to the manufacturers or builders within the 300 km radius shall be borne by the respective power plants and beyond 300 km, the cost shall be equally shared between the users and power plants. This notification also prescribes power stations to upload the details of stock available with them and keep updated regularly [15].

The latest notification issued on April 22, 2021, is focused on the utilization of legacy ash. According to the notification, thermal power plants need to utilize the legacy ash within 10 years from April 1, 2022 as stated in the notification [16]. The percentage utilization of legacy ash shall be based on the annual fly ah production, that is, the utilization of legacy ash should be at least 20% within 1st year; 35% within 2nd and 50% above 3rd to 10th year, from the issuance of the notification, failing to this will impose penalty as per quantity of unutilized legacy ash [16].

2.4 Fly ash utilization

Research scientists all around the world strive hard for the development of various modes of fly ash utilization in cost-effective and ecofriendly manner. Due to their immense efforts, the fly ash utilization increased steadily from the ~6.6 million tonnes (1996–1997) up to the ~215 million tonnes (2020–2021) [10]. **Figure 4** shows the details of the fly ash utilization in India during the year 2020–2021.

Figure 4 reveals that the considerable amount of fly ash is utilized by the cement sector, which is 25.8% of the total fly generated during the year. Then, other significant amount 15.6% of fly ash was utilized for the reclamation of low-lying area, followed by the sector of roads and flyovers, where around 15.0% of fly ash utilized. Bricks and tiles manufacturer have used about 12.9% of fly ash in 2020–2021, while ash dykes raising and filling of mining area utilized about 6.2 and 0.83%, respectively, of fly ash generated. Only 0.03% fly ash was utilized in the sector of agriculture and hydro-power sector. The quantity of fly ash remains unutilized during the year 2020–2021 was 7.59%, which is around ~108 million tonnes [10].

Restogi et al. have studied the potential to utilize fly ash in the field of different construction area and came out with the results for the fly ash utilization with the projected level until 2030. The study revealed that concrete and cement sectors have reached their threshold limit with accommodating around 35 to 40% of total fly ash generation in the forthcoming years, while the sectors of bricks and blocks are found to be the most potential mode for the fly ash utilization, with the capacity to

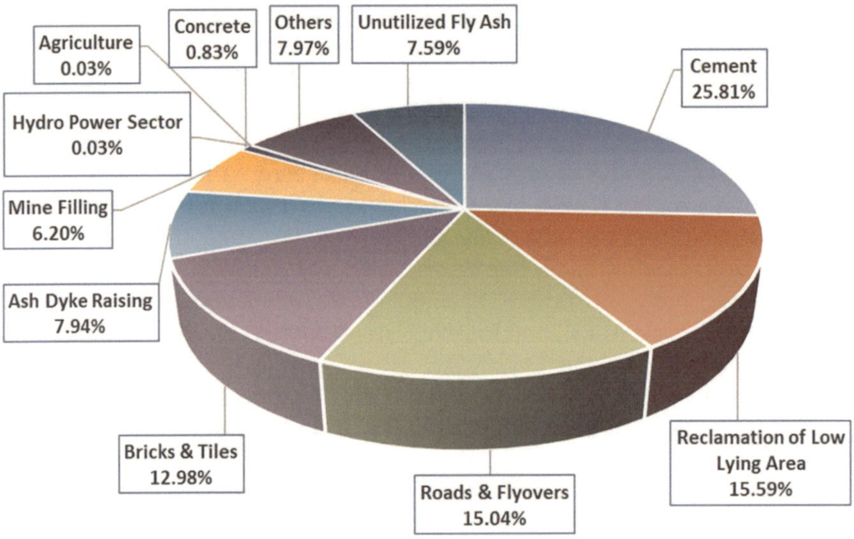

Figure 4.
Fly ash utilization in India during the year 2020–2021 as per CEA report, Aug-2021, Ministry of Power, INDIA [10].

accommodate all unutilized fly ash remaining during the year along with the potential to utilize the legacy ash within the next 15 to 20 years [17].

Though sectors such as mine filing (6.2%), reclamation of low-lying area (15.6%), ash dykes raising (7.9%), and roads and flyover (15%) consume ample amount of fly ash, it should be consider as the last option as these are still existence of threats to the environment after utilization of these modes. Hence, there is still huge quantity of fly ash available to be explored for its efficient utilization in such a manner so as to remove or reduce their harmful effects to the environment. These modes of utilization are followed by the manufacturer as the regulations have been enforced to utilize fly ash in these sectors as compulsion. If the mode of utilization that utilizes the fly ash as raw materials to produce value-added materials, then it will be utilized efficiently by the various industrialists to produce the value-added products, such as the precursor for the ceramics industries and development of ceramic membrane, for the development of adsorbents to remove toxic metals, drug impurities from the water, and development of catalysts such as zeolites for various organic transformations like methane to olefin conversion, biodiesel production, fast pyrolysis of Jatropha waste [18–25]. As fly ash consists of various elemental oxides, specifically SiO_2 and Al_2O_3, which are useful as precursors for the zeolites. Hence, intense research works have been carried out for the development of zeolites from the waste CFA [18, 26].

2.5 Properties of fly ash

The fly ash can be classified as two different types of classes (Class F and Class C) according to the ASTM C-618-3 standards, which are shown in the table below [27]:

- Class F: The fly ash produced after burning of the anthracite and bituminous type of coals. This kind of fly ash contains minimum 70% of silicon dioxide (SiO_2) plus aluminum oxide (Al_2O_3) plus iron oxide (Fe_2O_3) along with CaO less than 10% and possesses pozzolanic properties.

- Class C: The fly ash that produced after burning of lignite or sub-bituminous type of coals. This type of fly ash contains minimum 50% of silicon dioxide (SiO_2) plus aluminum oxide (Al_2O_3) plus iron oxide (Fe_2O_3) along with lime (CaO) content more than 10% (mostly found in the range of 15 to 20%, sometimes up to 40%) and possess cementitious properties along with pozzolanic properties [27–32].

The Indian coal fly ashes with low lime content are relatively higher in concentration of oxides of silica and alumina, whereas oxides of iron contents are found lower. Due to such properties, these fly ashes require higher temperature for fusion because at lower temperature the chances of glass formation are also low. In these types of fly ashes, the silica content is found almost double than the content of alumina, whereas in high-calcium fly ashes, the oxides of silica and alumina are found almost close to each other along with the significant amount of oxides of iron with respect to the low-lime fly ashes [30]. The heterogeneity studies of the fly ashes by various methods such as sieving, sink-float, and magnetic separation suggests that the heterogeneity in fly ashes with high lime contents is found higher as more variations in the compositions are observed in such kind of fly ashes [30, 33]. The general composition of CFA from various sources is summarized in **Table 1**.

Based on physical properties of the fly ash, it could be considered as fine glass powder with the regular spherical particles sizes in range of 0.5 μm to 100 μm. Such spherical particles shapes allow them to flow and blend freely in the mixture of concrete or cement. The fly ash particles possess ball bearing effect, which provide lubricating actions in the concrete with the plastic state and thus help to decrease the dry shrinkages [11, 34]. The concrete blended with fly ash tends to resist attack of water, sea water, mild acids, and sulfates, which ultimately increases the durability of the concrete mixture and makes it suitable for coastal environment [35, 36]. The fly ash having long-lasting pozzolanic properties, which is useful to tie up free lime, thus reduces the bleed channels by decreasing the permeability of the concrete. Such properties are also useful for increasing the structural strength of the concrete mixer over a time. Due to such kind of physical properties, around 27% of the fly ashes are utilized in the concrete and cement sector [34]. But, it is believed to be at the saturation level to be more utilized in this sector and needs to be explored the utilization of fly ash in other sectors, which turns it into value-added materials.

3. Fly ash utilization for the synthesis of zeolites

3.1 Reported synthesis protocols for fly ash-derived zeolites

3.1.1 Hydrothermal synthesis method

The zeolite formation most commonly proceeds through the hydrothermal synthesis process. The crystallizations or formation of zeolites proceeds through the hydrothermal synthesis process. The various parameters, such as temperature of crystallization, pH of gel, concentration of alkaline cations, time of crystallization, reaction conditions such as static or continuous stirring mode and autogenous pressure involved during the hydrothermal process, may define the formation of specific

SiO_2	Al_2O_3	Fe_2O_3	CaO	MgO	K_2O	TiO_2	P_2O_5	MnO	SO_3	Na_2O	Others	Ref.
CFA from Thermal Power Plant India (Class F)												
52.4	35.8	6.49	1.21	0.46	1.29	1.68	0.29	0.14	0.16	NR	0.06	[18]
South African CFA from Arnot coal power plant, Mpumalanga province (Class F)												
59.0	28.0	5.00	4.00	1.20	0.70	1.70	0.40	NR	NR	NR	0.00	[21]
Lignite CFA collected from Serbian Thermal Power Plant (Class F)												
59.2	22.6	6.28	7.48	2.19	1.60	NR	NR	NR	NR	0.24	0.44	[23, 24]
CFA collected from pulp and paper industries in Thailand (Class C)												
41.4	20.0	8.40	14.10	2.30	2.00	0.60	0.40	NR	4.90	0.40	5.50	[25]
CFA produced by coal gasification process, China (Class F)												
51.9	18.4	7.94	7.52	3.50	2.42	1.04	NR	NR	NR	1.95	5.39	[30]
CFA samples were collected from NTPC thermal power plant, West Bengal (India) (Class F)												
47.5	27.8	5.00	1.01	NR	11.44	1.39	NR	NR	NR	NR	5.89	[31]
CFA was obtained from a power plant, China (Class F)												
36.9	26.6	12.82	6.27	2.90	0.47	4.42	0.07	0.10	0.67	8.33	0.37	[32]
CFA of circulating fluidized bed combustion, from the Pingshuo Coal Gangue Power Plant, China (Class F)												
40.4	34.5	4.34	7.64	1.33	0.62	1.45	NR	NR	4.28	NR	5.44	[33]
CFA was from the Anhui Province of China (Class F)												
54.6	32.2	3.20	1.50	0.60	0.80	1.30	NR	NR	NR	0.20	5.60	[34]
CFA was from Thermal Power Plants in Bulgaria, TPP "AES Galabovo" (Class F)												
62.1	26.6	3.87	1.97	1.40	1.18	1.73	NR	NR	0.49	0.57	0.02	[35]
CFA was collected from Mae Moh Power Plant, Lumpang provice, Thailand (Class C)												
22.1	12.1	15.00	32.60	1.99	1.78	0.38	0.38	0.19	11.2	1.82	0.46	[36]
CFA was collected from Duvha Thermal Power Plant, Mpumalanga, South Africa (Class F)												
46.9	31.3	2.96	4.33	0.6	0.72	1.54	0.47	0.03	0.40	NR	10.7	[37, 38]
CFA obtained from the Midong Electric Power Company Limited (China) (Class F)												
43.4	19.8	NR	NR	NR	NR	NR	NR	NR	NR	NR	36.8	[39]
CFA used for this experiment was procured from Adhunik Power & Natural Resources Ltd., Jharkhand (Class F)												
68.9	10.1	9.83	1.11	2.07	0.67	1.05	NR	NR	NR	6.12	0.04	[40]
CFA from was collected from Figueira Power Plant, Parana State (PR), Brazil (Class F)												
40.9	13.5	27.2	2.60	0.31	5.00	2.60	1.40	0.11	1.40	NR	4.98	[41]
CFA was taken from a coal-fired power plant in Anhui Province, China (Class F)												
55.8	27.7	2.92	0.88	0.40	1.23	NR	NR	NR	NR	0.66	10.4	[42]

Unit of all values: %, CFA: Coal Fly Ash, NR: Not Reported, Ref.: References.

Table 1.
Chemical composition of coal fly ash (CFA) obtained from various sources.

types of zeolites with their unique properties. The gel composition used to crystallize zeolites mainly contains the sources of hetero elements (tetravalent or trivalent elements), mainly silica and alumina, which, further, takes part in the formation of framework structure of zeolites. The gel also comprises of the sources of mineralizing agents such as OH^- and F^- along with inorganic cations, organic species, and solvent (generally water), which solubilizes the reactive elements in the gel and enables them to transfer into the growth of the zeolite crystals [37].

In hydrothermal synthesis of zeolite, homogenized well-mixed sol-gel is prepared by mixing precursors of silica and alumina, source of inorganic cations, and mineralizing agent (i.e., NaOH or KOH) along with the water as a solvent. The gel is then transferred to a teflon-coated autoclave reactor (static or agitated) and then heated for specific period of time with or without stirring [18, 37, 38]. As the hydrothermal crystallization proceeds the high temperature and autogenously generated high-pressure environment triggered, the process of crystallization and the crystal growth proceeds according to the available environment to produce zeolite product. The parameters affecting the crystallization are the concentration of alkali, temperature of crystallization, time of aging, and liquid-to-solid ratio in the reactor. If the alkalinity is too low in the gel, there may be not enough Na^+ availability too trigger the crystallinity, while, if the alkalinity is high, the dissolution of the crystal nucleus would be accelerated, which further affects the crystallization of the zeolitic product [39, 40]. The aging plays an important role in the crystallization process. Aging promotes the formation of oligosilicate ions by the polymerization of the single silicate ions available in to the sol gel, which is further utilized for the crystallization of the zeolite. The temperature of the hydrothermal synthesis is another important parameter for the efficient crystal growth of zeolite. The autogeneous pressure is dependent on the temperature and volume of the gel mixture of the autoclave reactor. High temperature of the reactor triggers the autogeneous pressure in the reaction vessels and thus accelerates the speed of the crystallization of zeolites. Different kinds of zeolite crystals are observed at different temperatures. If the temperature is too low during the process, the pressure will be less, thereby leading to reduce crystallization speed and difficulty in the formation of targeted zeolite crystals [41, 42]. Chang and Shih have synthesized zeolite A and faujasite from Class F type fly ash at low temperature (38 and 60°C, respectively), but required longer reaction time (more than 3 days) for the crystallization of Zeolite A, while at 90°C Zeolite P formation is observed [43]. On the other hand, Yoshida and Inoue have synthesized the similar zeolite at 90°C but it is observed that at the higher reaction temperature, phase of Zeolite A started to disappear and formation of Zeolite P is observed [44]. Amoni et al. have synthesized zeolite A from CFA *via* hydrothermal process after applying magnetic treatment to remove magnetic fraction and acid treatment to remove other unwanted elements [45]. Kumar and Jena have synthesized highly crystalline zeolites with pure phase of zeolites Na-P1, hydroxy sodalite, and analcime zeolites by direct hydrothermal synthesis method, with and without homogenization of the gel precursors, at 150°C temperature and 60 h (to obtain Na-P1) and 20 h (to obtain hydroxy sodalite with homogenization and Analcime without homogenization) [46].

The liquid-to-solid ratio (LSR) should be maintained in accordance with the reactor capacity to achieve desired product. The optimum LSR is useful to generate desired autogeneous pressure, which ultimately triggered the crystallization process. It is also useful to maintain Na^+ concentration and the alkalinity of the gel composition. Generally, the optimum LSR is 66% to get enough autogeneous pressure during the hydrothermal process [40, 47]. The optimum aging time helps to produce stable silicon aluminum gel polymer, which further helps to get well-crystallized zeolite and also helps to reduce the time of crystallization. On the contrary, longer aging period may dissolve the colloidal skeleton, which affects the quality of crystallinity of the zeolites [48]. The hydrothermal synthesis methods suffer with few disadvantages such as it consumes high energy, longer reactions time, and sometimes suffering from lower crystallization efficiency [38]. Therefore, there is a need to explore other green synthesis approaches and pretreatment methods to prepare sol-gel materials before hydrothermal synthesis. The hydrothermal synthesis process is mostly followed for the zeolite crystallization, but to get targeted materials few pretreatments of the raw materials are necessary to improve the crystallization of pure zeolites.

Before the hydrothermal synthesis process, the pretreatment of the raw materials (specifically when waste CFA or other kind of waste raw materials is used as silica and alumina sources) favors the formation of the desired zeolitic materials. Such pretreatment methods involve the steps such as acid treatment of the waste CFA to remove undesired elements, that is, Fe_2O_3, CaO, MgO, K_2O, and Na_2O [49–51].

3.1.2 Alkali fusion method

The alkali fusion of the raw materials is another pretreatment method involving the fusion of the waste materials to extract maximum silica from it, which is further utilized to prepare gel composition to synthesize the zeolites *via* hydrothermal synthesis method. Fly ash is utilized as the sources of silica and alumina, which are the important starting ingredients for the synthesis of alumino-silicate or zeolite. Shigemoto et al. have introduced the pre-alkali fusion treatment prior to the hydrothermal synthesis process, to convert entire CFA particles into the zeolite [52]. The silica and alumina are present in the fly ash as the quartz and mullite phases. These phases of silica and alumina are less reactive as they are not easily soluble in the water and hence required suitable activation treatment, which weakens the bonds of silica and alumina with the oxygen available in the oxides of silica and alumina present in the fly ash [52, 53]. By applying the alkali fusion treatment, Shigemoto et al. have synthesized zeolite Na-X with 62% yield, while formation of zeolite Na-A was observed with the CFA enriched with the aluminum content [52]. Alkali fusion treatment is the method of choice for such kind of activation of fly ash, which extracts water soluble silicates and aluminates from the fly ash which further improve the crystallization of desired zeolites. Alkali fusion is carried out using NaOH and KOH classically [53, 54]. Chang et al. have established the method of alkali fusion followed by hydrothermal process as a general method to synthesize specific type of zeolites from the various fly ash sources [43]. They have converted the quartz and mullite phases of fly ashes into zeolites by melting fly ash with NaOH at 550°C. Their study reveals that the alkali fusion method followed by general hydrothermal method provides better crystallization then the simple hydrothermal process carried out without alkali fusion treatment [42, 43]. Murukutti and Jena have reported the alkali fusion treatment of fly ash that is carried out by Na_2CO_3 instead of NaOH or KOH. The use of Na_2CO_3 is cost effective with respect to the NaOH. The study reveal that when the fusion of fly ash is carried out with Na_2CO_3 or NaOH at 800°C for 2 h, fly ash transform in to the nephiline (under saturated alumino silicate), which further facilitates the crystallization of zeolites (Zeolite A at 100°C for 6 h and Zeolite X at 100°C for 8 h) through hydrothermal synthesis process, while cancrinite zeolite is formed at 100°C, with prolonged (48 h) time of hydrothermal process [53]. Park et al. have synthesized zeolite X and Y under molten condition with alkali without addition of water. The study reveal that the crystallization could not be completely accomplished using the method of molten salt due to low temperature and insufficient contact of fly ash with the alkali (NaOH). Hence, the product formed using molten salt method was found to be irregular in their morphological shapes, which could not be identical match with the particular zeolites. The polycrystals of zeolites were very well developed when hydrothermal synthesis methods were used as compared to molten salt method. The study reveals that the alkali fusion method followed by the hydrothermal treatment is much reliable in order to obtain the zeolites selectively from fly ash [42, 55].

Though the alkaline fusion treatment followed by hydrothermal process improves the crystallization and yield of the zeolites, it also suffers with few disadvantages. Such as, when the oxides of silica and alumina are not in a sufficient amount, during the alkali fusion of fly ash the oxides of the elements other than Si and Al also mixed with the alkali melt products. These elements are hard to separate from the alkali melt product mix, which further carried out in the hydrothermal process and mixed up with the zeolite products, ultimately affect the quality of zeolite crystals. Thus, the alkali fusion method is suitable only when fly ash contains higher amount of oxides of silica and alumina. Moreover, alkali fusion treatment required higher energy consumption [56, 57].

The hydrothermal synthesis method suffers with the problems of longer reaction time and high energy requirement for the activation of Mullite and glassy phase of fly ash. Similarly, alkali fusion methods also required high temperature from 550 to 850°C [43, 53]. Even though these methods have been utilized extensively to produce highly crystallized and pure form of zeolites, however, these methods are not economically or environmentally benign as it required high temperature along with the longer time period [58]. Moreover, constructions of the large scale furnace and its operations are too expensive, it also generate solid waste and the handling of large quantity of liquid waste generated through this process is also a matter of concern as it contains most of the toxic metals dissolved from fly ash [59]. Hence, there is a need to find greener synthesis methods which are economically viable and environmentally benign. Few efforts have been reported via microwave and ultrasound assisted synthesis methods of CFA based zeolites.

3.1.3 Microwave-assisted method

Querol et al. [39] have introduced the microwave assisted method for the crystallization of zeolite from waste CFA. The quality, yield and types of the zeolites obtained using microwave assisted synthesis method and traditional experimental process have been found to be quite similar, but the activation time in the microwave assisted synthesis has been drastically reduced to 30 min instead of 24–48 h in traditional methods [39, 42]. Inada et al. have studied the mechanism of the microwave assisted synthesis and found that appropriate microwave radiation could increase the rate of crystallization, on the contrary, prolonged radiation time could inhibit the crystal formation of the zeolites [60]. The microwave assisted method provides uniform nucleation in the supersaturated gel solutions with rapid crystallization [56]. LTA zeolite has been synthesized from CFA by Behin and co-workers using microwave assisted method at low power of microwave radiation (100 to 300 W) within 10 to 30 min of shorter radiation time [61]. Makgabutlane et al. have observed that the zeolite crystallization from fly ash could not be accomplished successfully using microwave assisted synthesis, it needs to be combined with the conventional methods like alkali fusion and hydrothermal synthesis process [58].

3.1.4 Ultrasound method

As the microwave assisted synthesis method is under development stage and the traditional methods to extract the silica and alumina from fly ash suffers from severe disadvantages such as high energy consumption, longer reaction time and high temperature of fusion (550 to 800°C), the ultrasound method is also explored as alternative methods, recently, for the activation of fly ash and further crystallization of zeolites [56, 62, 63]. Ju et al. have observed that the ultrasonic waves helps to improve the efficiency of the extraction of silica from fly ash with compare to traditional methods [63]. Ojumu and co-worker have reported the ultrasonic treatment method for the activation of fly ash instead of high energy consuming method of alkali fusion method followed by hydrothermal synthesis method to produce zeolite A with a low energy requirement and shorter reaction time [64]. The synthesis induced by the ultrasonic waves shows rapid zeolite formation with significant reduced temperature of crystallization than the conventional hydrothermal synthesis process [62].

Chen et al. have reported the microwave and ultrasound collaborative activation method, which significantly improve the activation efficiency of silica and alumina from the fly ash with energy friendly treatment method [56]. The novel EU-12 nanozeolite has been synthesized from coal fly ash by the ultrasonication treatment method followed by hydrothermal synthesis process with ~76% crystallinity [65].

3.1.5 Seed assisted synthesis method

Seed assisted synthesis method is also now a days useful to get desired zeolites using the hydrothermal synthesis process [18]. Seeding is employed in so many industrial crystallization processes to improve the quality of products. Seeding offers numerous advantages such as higher production yield, reduced induction period and contributing towards controlled particle size distribution during crystallization [37]. The details regarding the effect of seeding at the molecular level are still unexplored and need more systematic study in this area. Itabashi and coworkers have studied the effect of the zeolite seeds having different types of framework structures on the crystallization process to form targeted zeolites [66]. They have suggested a working hypothesis for the seed assisted crystallization of targeted zeolites based on the composite building units (cbu) of targeted zeolites, zeolites utilized as seeding agent and unseeded sol-gels. To validate the hypothesis, synthesis of ECR-18 has been carried out using the sol-gel composition, yielded Linde W without seed, but using the calcined ECD-18 as a seeding agent in this gel, formation of pure phase of ECD-18 is observed [66]. Similarly, zeolite CIT-1 (CON type zeolite) has been synthesized using Beta type zeolite as a promising seeding agent [67].

During the synthesis of zeolite through various steps, aging is the step where the silicates and aluminates from the sol-gel mixture converts into the precursor for the crystallization of zeolites. The crystallization of zeolite then proceeds through two steps, nucleation and crystal growth. The steps involving the formation of precursor before the starting of nucleation possess high activation energies and hence considered as the induction period. The presence of seeding agent can alter the nucleation process. The crystal of seeding zeolites act as nuclei and the crystals of targeted zeolites start to grow on the active surfaces of the seeding agent. As the crystals of products started to grow, the activation energies are considered to be lower and the growth of the crystals is faster than the nucleation [37]. The equilibrium between secondary nucleation and growth of seed crystals depends upon the quantity of gel materials, nature of the system and degree of agitation. If the seed crystals with sufficient surface area which can accommodate maximum available flux of growth species by preventing the effective solution to reach high levels of super saturation. In this condition, the most of the growth of the crystals in the system will take place on the available active area of the seeded crystals. In such case, the size and rate of growth can be controlled and predicted closely by kinetic study. If the seeded crystals with reduced surface area is added to the solution, then the natural super saturation and self-nucleation will no longer be suppressed and ultimately lead to the different kind of crystal size and type than targeted zeolites [37, 68–70].

The seed assisted methods is also explored for the OSDA (Organic Structure Directing Agent) free and ultrafast synthesis of specific type of zeolites. The zeolites obtained through OSDA free seed assisted synthesis are found well crystallized with less defects within the structure of the products than the zeolites synthesized with the OSDA. This is because the seed crystal accelerate the crystal growth and there is no need of calcination to remove OSDA, which often leads to the dealumination and resulting in the creation of defective sites [37]. Liu and coworkers have reported the ultrafast seed assisted synthesis method for the synthesis of AlPO-5 and SSZ-13, the method offers rapid crystallization of zeolites AlPO-5 (1 min) and SSZ-13 (10 min). The crystallization is carried out in tubular stainless steel reactor with fast heating in preheated oil bath. This kind of reactor can be useful for mass production with continuous operation [37, 71, 72].

Different methods that are used for the synthesis of CFA-derived zeolites along with the fly ash elemental compositions are summarized in **Table 2**.

CFA source and chemical composition	Synthesis/pretreatment methods: Reaction conditions and types of synthesized zeolites	Ref.
CFA from thermal power plant, India **(Class F)** **Chemical Composition:** SiO_2: 52.4%, Al_2O_3: 35.8% Fe_2O_3: 6.5%, CaO: 1.2%	**Aging:** Room temperature for 1 h **Seed Assisted Hydrothermal:** 170°C, 24 h, Agitated **Seeding Agent:** 5% Mordenite (with respect to silica) **Gel composition ratio:** $6Na_2O:Al_2O_3:30SiO_2:780H_2O$ **Zeolite:** Mordenite	[18]
South African CFA from Arnot coal power plant, Mpumalanga province **(Class F)** **Chemical Composition:** SiO_2: 59.0%, Al_2O_3: 28.0% Fe_2O_3: 5.0%, CaO: 4.0%	**Acid Treatment:** Conc. H_2SO_4 (extraction followed by filtration for the removal metals like Al and Fe) **Aging:** Room temperature for 2 h. **Hydrothermal:** 160°C, 72 h **Gel composition ratio:** 5.8Si: Al:1.7Na:0.9TPABr:306.9H_2O **Zeolite:** ZSM-5	[21]
CFA was obtained from the energy sector, Poland **(Class F)** **Chemical Composition:** NR	**Alkali Fusion:** with NaOH, Ratio of FA to NaOH: 0.8, Temperature: 700°C **Aging/Hydrothermal:** 700°C temp. for 6 h, static **Impregnation of Ni:** 10–15 wt% Nickel Nitrate addition at room temperature in zeolite X **Drying and calcined:** drying overnight at 105°C and calcined at 500°C for 5 h. **Zeolite:** Ni/Zeolite X	[20]
Lignite CFA collected from Serbian Thermal Power Plant **(Class F)** **Chemical Composition:**	**Acid Treatment:** 6 M HCl at 80°C for 6 h, solid-to-liquid ratio of 1:5 **Hydrothermal:** 260°C temperature for 4 h with 6.25M NaOH solution, Agitated **ES900 preparation:** Egg Shells were dried in the air for	[23, 24]

CFA source and chemical composition	Synthesis/pretreatment methods: Reaction conditions and types of synthesized zeolites	Ref.
SiO_2: 59.2%, Al_2O_3: 22.6% Fe_2O_3: 6.3%, CaO: 7.5%	a few days, ground into powder, calcined at 900°C at 5° C/min for 4 h. **Cao-Cancrinite preparation:** ES900 (5–10 wt%) added to 10 wt% Zeolite in alcoholic suspension and sonicated using low frequency ultrasound device for 15 min. Evaporate alcohol and drying the material overnight at 110°C, calcined at 450 to 600°C at 5°C/min heating rate for 4 h. **Zeolite:** CaO-Cancrinite zeolite	
CFA was collected from pulp and paper industries in Thailand **(Class C)** **Chemical Composition:** SiO_2: 41.4%, Al_2O_3: 20.0% Fe_2O_3: 8.4%, CaO: 14.1%	**Aging:** at 90°C for 3 h (with 1-3M NaOH solution) **Gel SiO_2/Al_2O_3 molar ratio:** 23.9 **OSDA:** Tetrapropylammonium bromide (TPABr) **Hydrothermal:** 160°C temperature for 24 h and 72 h **Dried and calcined:** dried at 100°C overnight and calcined at 540°C for 5 h to remove OSDA **Zeolite:** ZSM-5	[25]
CFA produced by coal gasification process, China **(Class F)** **Chemical Composition:** SiO_2: 51.9%, Al_2O_3: 18.4% Fe_2O_3: 7.9%, CaO: 7.5%	**Acid Treatment:** 10M HCl (Extraction followed by filtration to remove alkaline oxides, i.e., Fe_2O_3, CaO, Na_2O, K_2O and MgO.) **Alkali Fusion:** with NaOH for 2 h, Ratio of FA to NaOH: 0.8–1.4 g/g, Temperature: 400-700°C **Aging:** Room temperature for 12 h with stirring, after mixing with Dist. H_2O (5 mL/g) **Hydrothermal:** 90-110°C, 12 h, Static **Zeolite:** Zeolite Na-P1	[30]
CFA was obtained from a power plant in Datong City (Shanxi province, China). **(Class F)** **Chemical Composition:** NR	**Hydrothermal:** 95°C, 24 h **Impregnation of La:** Adding 0.23M $LaCl_3$ to the zeolite solution, stirring at 25°C for 4 h **Zeolite:** Modified Zeolite type Na–P1 (La-FA)	[71]
CFA was collected from NTPC thermal power plant, West Bengal, India **(Class F)** **Chemical Composition:** SiO_2: 47.5%, Al_2O_3: 27.8% Fe_2O_3: 5.0%, CaO: 1.01%	**Acid Treatment:** Conc. H_2SO_4 (Extraction followed by filtration for the removal metals like Al and Fe) **Alkali Fusion:** with Na_2CO_3 for 2 h, Ratio of FA to Na_2CO_3: 0.8–2.0 g/g, Temperature: 800°C **Aging/Homogenization:** Room temperature for 2 h with 3M NaOH. **Hydrothermal:** 100°C, 6 h, 8 h and 48 h **Gel composition ratio:** 88Na:88Al:104Si:384O:194H_2O (Zeolite-X), 12Na:12Al:12Si:48O:27H_2O (Zeolite-A) **Zeolite:** Zeolite-A and Zeolite-X	[31]
CFA was collected from NTPC thermal power plant, West Bengal, India **(Class F)** **Chemical Composition:** SiO_2: 47.5%, Al_2O_3: 27.8%	**Aging:** Room temperature for 48 h with 1M NaOH (LSR: 10:1) for Na-P1, Room temperature for 48 h with 5M NaOH (LSR: 10:1) for hydroxy sodalite, Room temperature for 48 h with 1-3M NaOH (LSR: 10:1) for Analcime	[31, 72]
Fe_2O_3: 5.0%, CaO: 1.01%	**Hydrothermal:** 150°C, 10-60 h (Na-P1), 150°C, 10-20 h (Hydroxy Sodalite), 200°C, 10 h (Analcime), Static **Gel composition ratio:** 2.85SiO_2: Al_2O_3: xNa_2O:199H_2O (x = ~1.79–8.99) **Zeolite:** Zeolite Na–P1, Hydroxy Sodalite and Analcime	

CFA source and chemical composition	Synthesis/pretreatment methods: Reaction conditions and types of synthesized zeolites	Ref.
CFA was obtained from a power plant, China. **(Class F)** **Chemical Composition:** SiO_2: 36.9%, Al_2O_3: 26.7% Fe_2O_3: 12.8%, CaO: 6.3%	**Activation of FA:** Using Microwave and Ultrasound (MU) Combined Synthesis Instrument with 4M NaOH (LSR: 20, 10, 5), at 100°C for 60 min, power 0-1000 W, ultrasound frequency 25 KHz (Activated Liquid named as: 5-MU-L, 10-MU-L and 20-MU-L, respectively as per LSR: 5, 10, 20) **Hydrothermal using microwave radiation:** 100°C, 30-60 min **Gel composition ratio:** $1Al_2O_3$: $1.17SiO_2$: $2.90Na_2O$: $277.97H_2O$ (5-MU-L), $1Al_2O_3$: $1.17SiO_2$: $4.84Na_2O$: $463.93H_2O$ (10-MU-L), $1Al_2O_3$: $1.17SiO_2$: $8.91Na_2O$: $854.43H_2O$ (20-MU-L) **Zeolite:** LTA zeolites (10-MU-75-LTA)	[32]
CFA of circulating fluidized bed combustion, from the Pingshuo Coal Gangue Power Plant in Shanxi Province, China, **(Class F)** **Chemical Composition:** SiO_2: 40.4%, Al_2O_3: 34.5% Fe_2O_3: 4.3%, CaO: 7.6%	**Acid Treatment:** 4.41M HCl (Extraction followed by filtration for the removal $CaSO_4$, CaO, Fe_2O_3, and MgO) mixing for 2 h at 90°C, LSR: 5 mL/g **OSDA:** CTAB (0.5 and 1 g/g) **Hydrothermal:** 90, 110 and 130°C, 6, 9 and 12 h, Agitated **Zeolite:** Na-P1 zeolite	[33]
CFA was from the Anhui Province of China **(Class F)** **Chemical Composition:** SiO_2: 54.6%, Al_2O_3: 32.2% Fe_2O_3: 3.2%, CaO: 1.5%	**Hydrothermal:** 170°C, 3 h, with 0.5M NaOH, LSR: 15 mL/g, Agitated **Ca/Si molar ratio:** 0.8 **Zeolite:** Zeolites-calcium silicate hydrate composite	[34]
CFA was from Thermal Power Plants in Bulgaria, TPP "AES Galabovo" **(Class F)** **Chemical Composition:** SiO_2: 62.1%, Al_2O_3: 26.6% Fe_2O_3: 3.9%, CaO: 2.0%	**Alkali Fusion:** with NaOH for 1 h, Ratio of FA to NaOH: 0.5 g/g, Temperature: 550°C **Additives:** NaCl, KCl, Na_2SiO_3 and Na_2CO_3 **Sonication:** Residue in Dist. H_2O and sonication for 15-48 min **Aging:** Room temperature for 4 h **Hydrothermal:** 90°C, 4-8 h **Zeolite:** Nanozeolite Na-X (ZFH2)	[35]
CFA was collected from Mae Moh Power Plant, Lumpang province, Thailand **(Class C)** **Chemical Composition:** SiO_2: 22.1%, Al_2O_3: 12.1% Fe_2O_3: 15.0%, CaO: 32.6%	**Acid Treatment:** 10% HCl, LSR: 25 mL/g, heated at 80°C for 1h with stirring at 300 rpm **Alkali Fusion:** with NaOH for 2 h, Ratio of FA to NaOH: 0.8–1.4 g/g, Temperature: 400–700°C **Aging:** Room temperature for 2 h **Silicate Solution (A):** FA:NaOH (1.67M) ratio: 10 mL/g, refluxed in R.B.F. at 80°C for 2h with stirring at 300 rpm **Aluminate Solution (B):** 2.4g $NaAlO_2$, 30mL NaOH (1.67M), stirred at 200 rpm for 30 min	[36]
	Aging: Solution A and B mixed on stirrer at 200 rpm for 30 min **Hydrothermal:** Mixed solution in nickel crucible and heated in hot air oven at 100°C for 340 min (5–6 h) **Zeolite:** Zeolite A (ZCF)	
CFA was collected from Duvha Thermal Power Plant, Mpumalanga, South Africa **(Class F)**	**Dissolution using Microwave radiation:** LSR (FA:2.5M NaOH): 7.5 mL/g, at 100°C, varied power (300, 600 and 900 W), and time (5, 10, and 15 min), Centrifuge to get slurry and extract.	[37, 38]

CFA source and chemical composition	Synthesis/pretreatment methods: Reaction conditions and types of synthesized zeolites	Ref.
Chemical Composition: SiO_2: 46.9%, Al_2O_3: 31.3% Fe_2O_3: 3.0%, CaO: 4.3%	**For Slurry** **Adjustment of Si/Al ratio:** with 28 mL $NaAlO_2$ Solution in 72 mL slurry (0.6 g $NaAlO_2$ + 1.2 g NaOH + 20 mL Dist. H_2O) **Hydrothermal using Microwave radiation:** at 300 and 600 W power, 10–20 min **For Extract** **Adjustment of Si/Al ratio:** with 28 mL $NaAlO_2$ Solution in 72 mL extract (0.6 g $NaAlO_2$ + 1.2 g NaOH + 20 mL Dist. H_2O) **Hydrothermal using Microwave radiation in:** at 300 and 600 W power, 10–20 min **Zeolite:** Zeolite A and Sodalite	
CFA was obtained from the Midong Electric Power Company Limited (China) **(Class F)** **Chemical Composition:** SiO_2: 43.4%, Al_2O_3: 19.8% Fe_2O_3: NR, CaO: NR	**Alkali Fusion:** with NaOH for 1 h, Ratio of FA to NaOH: 0.5 g/g, Temperature: 750°C, N_2 atmosphere **Aging:** Room temperature for 2 h (after ground and suspension of molten product in water) **Dried:** at 50°C for 12 h **Zeolite:** Carbon–Zeolite Composites (CZC)	[39]
CFA was taken from the Paiton PLTU, power plants in Indonesia **(Class C)** **Chemical Composition:** NR	**Dissolution:** LSR (FA:4M NaOH): 62 mL/g, stirring at 750 rpm at 100°C for 5 h **Adjustment of Si/Al ratio:** with 0.75g of $Al(NO_3)_3$ $9H_2O$ to mixture, stirring speed 750 rpm for 30 min **OSDA:** 2 g of TPABr in mixture, stirring under 100°C for 2 h **Aging:** Room temperature for 2 h. **Hydrothermal:** 160°C, 3–18 bar, 48 h **Dried and calcined:** dried at 110°C for 1 h and calcined at 550°C for 6 h to remove OSDA **Zeolite:** Zeolite obtained with Alumina was Analcime, Zeolite obtained without Alumina was ZSM-5	[73]
CFA was obtained from Adhunik Power & Natural Resources Ltd., Jharkhand. **(Class F)** **Chemical Composition:** SiO_2: 69.0%, Al_2O_3: 10.1% Fe_2O_3: 9.8%, CaO: 1.1%	**Calcination (FA):** 850°C, 90 min for the removal of unburnt carbon **Acid Treatment:** 3M HCl **Alkali Fusion:** with NaOH for 2 h, Ratio of FA to NaOH: 0.125 g/g **Ultrasonication:** at 700 W power, 20kHz, 60% amplitude for 10–30 min **Aging:** Room temperature for 24 h **Hydrothermal:** 100°C, 8 h, Static **Zeolite:** EU-12	[40]
CFA was collected from Figueira Power Plant, Parana State (PR), Brazil **(Class F)** **Chemical Composition:** SiO_2: 40.9%, Al_2O_3: 13.5% Fe_2O_3: 27.2%, CaO: 2.6%	**Hydrothermal:** 100°C, 24 h, with LSR: 8 mL/g (FA:3.5M NaOH) **Zeolite:** Hydroxy-sodalite	[41]
CFA was taken from a coal-fired power plant in Anhui Province, China **(Class F)** **Chemical Composition:** SiO_2: 55.8%, Al_2O_3: 27.7% Fe_2O_3: 2.9%, CaO: 0.9%	**Activation:** with Na_2CO_3 for 80 min, Ratio of FA to Na_2CO_3: 1 g/g, Temperature: 850°C, washed with water and filtered **Dissolution (Leaching of silicon and aluminum):** LSR (FA:2.5/3.75/**5.0** M NaOH): 1:10/15/**20** mL/g, stirring at **60**/75/90°C for 0.5/**1.0**/1.5 h and centrifuged	[42]

CFA source and chemical composition	Synthesis/pretreatment methods: Reaction conditions and types of synthesized zeolites	Ref.
	Synthesis of precipitated silica: Purging CO_2 gas into water washing solution till pH 8.00, aging at room temperature for 24 h in centrifugal tube and separate out solids, washed with water and dried at 100°C to get silica (precipitated) **Synthesis of Zeolite:** Aging of alkali leaching solution at room temperature for 24 h **Hydrothermal:** 100°C, 1/3/5 h **Zeolite:** Zeolite A and Sodalite	
CFA received from the Italian ENEL thermoelectric power plant in Cerano (Brindisi) and Fusina (Venice), Italy **(Class F)** **Chemical Composition:** SiO_2: 48.5%, Al_2O_3: 26.0% Fe_2O_3: 4.4%, CaO: 6.4%	**Alkali Fusion:** with NaOH for 1 h, Ratio of FA to NaOH: 0.83 g/g, Temperature: 550°C **Dissolution:** in artificial sea water **Hydrothermal:** incubated up to 192 h at 25°C **Zeolite:** Zeolite A (up to 96 h incubation) and Zeolite X (more than 96 h incubation)	[2, 49]

NR: Not Reported.

Table 2.
Methods for the synthesis of CFA-derived zeolites with CFA composition.

It is evident from the **Table 2** that, Zeolite Na-P, Zeolite X and Zeolite A are the most commonly synthesized zeolites from the CFA [2, 32, 41, 49, 51, 53, 56, 58, 73, 74]. It is observed that, at the lower temperature, at around, ~100°C, the conversion of CFA favors the formation of Zeolite Na-P and Zeolite A crystals [2, 32, 41, 49, 51, 53, 56, 73]. While at a high temperature, at around, ~160°C and a longer period of crystallization, CFA tends to form ZSM-5 crystals [21, 25, 75]. Formation of Analcime (in presence of high alumina contents in gel mixture) and Mordenite crystals are also observed at higher temperature of crystallization [18, 46, 53, 75]. When the significant amount of Na^+ cations are available in the gel composition, the crystallization of gel tends to form Analcime type zeolites, while at lower concentration, formation of Zeolite Na-P is observed [46, 49, 53, 75]. Zeolite A and X are crystallize even at room temperature in the presence of salt water (artificial seawater) but requires longer incubation period [2, 41]. Microwave assisted synthesis methods reduces the crystallization time for the formation of Zeolite A and Sodalite to only 10–60 min, which in conventional method takes up to few hours and sometimes up to few days [56, 58, 74]. While the use of ultrasonication during the crystallization of Nanozeolite-Na-X and EU-12 provides efficient dissolution of Si and Al within a few minutes compared to the conventional methods, thereby helping to reduce the period of hydrothermal process from few hours to only few minutes [56, 65].

4. Properties and applications of fly ash-derived zeolites

Zeolites exhibit unique properties such as higher thermal stability, high surface area and pore volume, chemical resistivity, and possess both acidic sites, Lewis and Brønsted. These properties enable zeolites to be utilized as catalysts and adsorbents in numerous industrial processes, such as cracking of petroleum crude, trans alkylation, hydro-isomerization, methylamine synthesis, and disproportionation [76]. Zeolites exhibit excellent cation exchange capacity making it capable to be utilized in the field of environmental remediation as the adsorbents and exchangers for the successful removal of pollutants such as toxic metals, nutrients and radionuclides. For example, as shown in **Table 2**, Kumar and Jena, 2022 have synthesized Zeolite NaP1 from the waste CFA, which possess good CEC (Cation Exchange Capacity), that is, 4.2 meq. NH_4^+/g. They have evaluate its performance for the removal of Sr^{2+} and Cs^{2+} from the nuclear waste and found 92.5 and 39.3 mg/g adsorption capacities, respectively [46]. Similarly, Vichaphund et al. have synthesized ZSM-5-type zeolite from the CFA shows acidity values of 0.979 mmol NH_4^+/g and further utilized it as catalyst for the fast pyrolysis of Jatropha waste. This catalyst shows high selectivity toward aromatic hydrocarbons formation and obtained with the yields of 97.4% [25]. The synthesized zeolites (from CFA) have been explored mainly as catalyst and sorbents. **Table 3** summarizes the properties and applications of fly ash-derived zeolites.

Table 3 reveals that the BET surface area (S_{BET}) of CFA-derived zeolites synthesized *via* classical methods was found to be less, which is in the range of 7.8–117 cm^2/g. The surface area of such zeolite can be improved by employing advance technique, such as ultrasonication and microwave irradiation combined with ultrasonication. It was observed that when ultrasonication method was utilized, the formation of nano-zeolite Na-X (ZFH2) was observed, this nano-zeolite possesses high surface area (486 cm^2/g) as shown in **Table 3** [77]. Similarly, EU-12 zeolite synthesized using sonication also shows high surface area (236 cm^2/g) [65]. The combined microwave and ultrasonication technique also seems to provide high surface area zeolites, such as

Fly ash source	Types of zeolites, properties and applications	Ref.
CFA from thermal power plant, India **(Class F)** **Properties:** S_{BET}: 2.9 cm^2/g Pore Volume: 0.009 cm^3/g Pore Size: 10.7 nm CEC: 0.32 meq Na^+/g	**Mordenite** Properties: S_{BET}: 117 cm^2/g Pore Volume: 0.20 cm^3/g Pore Size: 6.76 nm Acidity: 1.18 mmol NH_4^+/g CEC: 1.72 meq Na^+/g Application as Sorbent: For Pb^{2+} and Cd^{2+} removal. Removal Efficiency: 94.1% (Pb^{2+}), 88.2% (Cd^{2+}), Adsorption capacity: 20.2 mg/g (Pb^{2+}), 28.6 mg/g (Cd^{2+}) [at room temperature, Contact time: ~60 min, Dose: 10 g/L, pH: 5, Initial conc.: 0.004M (Pb^{2+}), 0.002M (Cd^{2+})]	[18]
South African CFA from Arnot coal power plant, Mpumalanga province **(Class F)** **Properties:** NR	**ZSM-5** Properties: S_{BET}: 328 cm^2/g Pore Volume: NR Pore Size: NR Acidity: 0.31 mmol H^+/g Application as Catalyst: For methanol to olefin (MTO) conversion Propylene selectivity: 35%, Light olefin (C_2–C_4) selectivity: 66%, [Temp.: 450°C, Weight hourly space velocity (WHSV): 1.12/h, Time: 1 h]	[21]

Fly ash source	Types of zeolites, properties and applications	Ref.
CFA was obtained from the energy sector, Poland **Properties:** NR	**Ni/zeolite X** Properties: S_{BET}: 190 cm^2/g Pore Volume: 0.149 cm^3/g Pore Size: 0.68 nm Acidity: 0.38 mmol H^+/g Application as Catalyst: For CO_2 methanation, CO_2 conversions: 53%, CH_4 selectivity: 90% [Temp.: 450°C, Gas hourly space velocity (GHSV): 12000/h]	[20]
Lignite CFA collected from Serbian Thermal Power Plant **(Class F)** **Properties:** S_{BET}: 2.1 cm^2/g Pore Volume: 0.003 cm^3/g Pore Size: NR	**CaO-Cancrinite zeolite** Properties: S_{BET}: 14.9 cm^2/g Pore Volume: 0.06 cm^3/g Pore Size: 13.5 nm Basicity: 0.92 mmol CO_2/g Application as Catalyst: For the biodiesel production, Triacylglycerols conversion: 96.5% [Temp.: 60°C, MTO ratio: 12:1, Catalyst Load: 4 wt%, Time: 2 h]	[23, 24]
CFA was collected from pulp and paper industries in Thailand **(Class C)** **Properties:** NR	**ZSM-5 (Synthesized with 3M NaOH at 72 h)** Properties: S_{BET}: 453 cm^2/g Pore Volume: 0.24 cm^3/g Pore Size: 3.31 nm Acidity: 0.979 mmol NH_4^+/g Application as Catalyst: For the fast pyrolysis of Jatropha waste, Aromatic Hydrocarbon Yield: 97.4%, Oxygenated Compound Yield: 0.8%, N-Containing Compound Yield: 1.7%, [Temp.: 500°C, Jatropha:catalyst ratio: 1:10]	[25]
CFA produced by coal gasification process, China **(Class F)** **Properties:** NR	**Zeolite Na-P1** Properties: S_{BET}: NR Pore Volume: NR Pore Size: NR CEC: 2.75 meq NH_4^+/g Application as Sorbent: For Cr^{6+} removal from water, Adsorption capacity: 17.9 mg/g [at 20°C, Contact time: 2 h, Dose: 5 g/L, pH: 3, Initial conc.: 100 mg/L]	[30]
CFA was obtained from a power plant in Datong City (Shanxi province, China). **(Class F)** **Properties:** S_{BET}: 5.5 cm^2/g Pore Volume: 0.01 cm^3/g Pore Size: NR	**Modified Zeolite type Na–P1 (La-FA)** Properties: S_{BET}: 59.9 cm^2/g Pore Volume: 0.11 cm^3/g Pore Size: 3.9 nm Application as Sorbent: For phosphorous removal from water Removal Efficiency: 72.8% Adsorption capacity: 23 mg/g (pH 4), 11 mg/g (pH 8.5), [at 25°C, Contact time: 30 h, Dose: 1.0 g/L, pH: 8.5, Initial conc.: 30 mg/L]	[71]

Fly ash source	Types of zeolites, properties and applications	Ref.
CFA samples were collected from NTPC thermal power plant, West Bengal (India) **(Class F)** **Properties:** S_{BET}: 4.04 cm^2/g Pore Volume: NR Pore Size: NR	**Zeolite-A (ZA) and Zeolite-X (ZX)** Properties: S_{BET}: 58.3 cm^2/g (ZA), 164.3 cm^2/g (ZX) Pore Volume: 0.071 cm^3/g (ZA), 0.054 cm^3/g (ZX) Pore Size: 5.108 nm (ZA), 4.537 nm (ZX) Application as Sorbent: For Sr^{2+} and Cs^{2+} removal from nuclear waste Removal Efficiency: >90% (Sr^{2+}), >50% (Cs^{2+}), Adsorption capacity: 54.1 mg/g (ZA), 53.1 mg/g (ZX) for Sr^{2+} and 95.7 mg/g (ZA), 93.1 mg/g (ZX) for Cs^{2+} [at 25°C, Contact time: 24 h, Dose: 10.0 g/L, pH: 7.0, Initial conc.: 1000 mg/L]	[31]
CFA samples were collected from NTPC thermal power plant, West Bengal (India) **(Class F)** S_{BET}: 4.04 cm^2/g Pore Volume: NR Pore Size: NR	**Zeolite Na–P1 (ZP), Hydroxy Sodalite (ZS) and Analcime (ZA)** Properties: S_{BET}: 63 cm^2/g (ZP), 17 cm^2/g (ZS), 8 cm^2/g (ZA) Pore Volume: NR Pore Size: 8 nm (ZP), 6 nm (ZS), NR (ZA) CEC (meq NH_4^+/g): 4.2 (ZP), 0.8 (ZS), 0.6 (ZA) Application as Sorbent: For Sr^{2+} and Cs^{2+} removal from nuclear waste Adsorption capacity: 92.5 mg/g (ZP), 43 mg/g (ZS), 11 mg/g (ZA) for Sr^{2+}, 39.3 mg/g (ZP), 15 mg/g (ZS), 0 mg/g (ZA) for Cs^{2+} [at 25°C, Contact time: 24 h, Dose: 10.0 g/L, pH: 7.0, Initial conc.: 1000 mg/L]	[31, 72]
CFA was obtained from a power plant, China. **(Class F)** **Properties:** NR	**LTA zeolites (10-MU-75-LTA)** Properties: S_{BET}: 442 cm^2/g Pore Volume: 0.30 cm^3/g Pore Size: 17-40 nm Acidity: 3.07 mmol NH_4^+/g Application as Catalyst: For the cracking of 1,3,5-tri-isopropylbenzene, Conversion: 25.40 wt% Benzene Selectivity: 11.05%, isopropyl-benzene Selectivity: 21.27%, 1,3-diisopropyl-benzene Selectivity: 16.58%, Propylene Selectivity: 51.1% [at Temp.: 300°C, Catalyst Load: 1.45 g/L]	[32]
CFA derived from circulating fluidized bed combustion, from the Pingshuo Coal Gangue Power Plant in Shanxi Province, China. **(Class F)** **Properties:** NR	**Na-P1 zeolite** Properties: S_{BET}: 43.96 cm^2/g Pore Volume: 0.0945 cm^3/g Pore Size: 8.21 nm Application as Sorbent: For Pb^{2+} removal from water Removal Efficiency: 99% (at 100-300 mg/L Pb^{2+}) Adsorption capacity: 425 mg/g [at 25°C, Contact time: 60 min, Dose: 0.2 g/L, pH: 6.0, Initial conc.: 500 mg/L]	[33]

Fly ash source	Types of zeolites, properties and applications	Ref.
CFA was from the Anhui Province of China **(Class F)** **Properties:** S_{BET}: 22.9 cm^2/g Pore Volume: 0.1 cm^3/g Pore Size: 8.1 nm	**Zeolites-calcium silicate hydrate composite** Properties: S_{BET}: 96.5 cm^2/g Pore Volume: 0.4 cm^3/g Pore Size: 9.1 nm Application as Sorbent: For Pb^{2+}, Ni^{2+}, Cd^{2+}, Zn^{2+}, Cu^{2+} and Cr^{3+} removal Adsorption capacity: In a single system: Pb^{2+}: 409.4 mg/g; Ni^{2+}: 222.4 mg/g; Cd^{2+}: 147.5 mg/g; Zn^{2+}: 93.2 mg/g; Cu^{2+}: 101.1 mg/g; Cr^{3+}: 157.0 mg/g In a multiple system: Pb^{2+}: 121.9 mg/g; Ni^{2+}: 4.6 mg/g; Cd^{2+}: 70.8 mg/g; Zn^{2+}: 27.5 mg/g; Cu^{2+}: 34.4 mg/g; Cr^{3+}: 9.7 mg/g [at 25°C, Contact time: 120 min, Dose: 0.5 g/L, pH: 4.5, Initial conc.: 25-400 mg/L]	[34]
CFA was collected from Mae Moh Power Plant, Lumpang province, Thailand **(Class C)** **Properties:** S_{BET}: 1.390 cm^2/g Pore Volume: 0.006 cm^3/g Pore Size: 7.97 nm	**Zeolite A (ZCF)** Properties: S_{BET}: 10.440 cm^2/g Pore Volume: 0.030 cm^3/g Pore Size: 5.57 nm Application as Sorbent: For Pb^{2+} removal from water Removal Efficiency: 100% Adsorption capacity: 556 mg/g [at 25°C, Contact time: 1 h, Dose: 0.1 g/L, pH: 5.0, Initial conc.: 50 mg/L]	[36]
CFA was obtained from the Midong Electric Power Company Limited (China) **(Class F)** **Properties:** S_{BET}: 11.0 cm^2/g Pore Volume: 0.03 cm^3/g Pore Size: 3.4 nm	**Carbon–Zeolite Composites (CZC)** Properties: S_{BET}: 62.5 cm^2/g Pore Volume: 0.31 cm^3/g Pore Size: 12.7 nm Application as Sorbent: For Pb^{2+} removal from water Adsorption capacity: 186 mg/g [at room temperature, Contact time: 40 min, Dose: 10 g/L, pH: 7.0, Initial conc.: 500 mg/L]	[39]
CFA was obtained from Adhunik Power & Natural Resources Ltd., Jharkhand. **(Class F)** **Properties:** NR	**EU-12** Properties: S_{BET}: 236.03 cm^2/g Pore Volume: 0.273 cm^3/g Application as Sorbent: For Rhodamine B (RB) dye removal. Removal Efficiency: 67.32% [at room temperature, Contact time: 150 min, Dose: 1 g/L, pH: 3.0, Initial conc.: 10 mg/L]	[40]
CFA was collected from Figueira Power Plant, Parana State (PR), Brazil **(Class F)** **Properties:** S_{BET}: 12.47 cm^2/g Pore Volume: NR Pore Size: NR	**Hydroxy-sodalite** Properties: S_{BET}: 66.12 cm^2/g Pore Volume: NR Pore Size: NR	[41]

Fly ash source	Types of zeolites, properties and applications	Ref.
	Application as Sorbent: For Zn^{2+} and Cd^{2+} removal Removal Efficiency: 95% (Zn^{2+}), 97% (Cd^{2+}) [at room temperature, Contact time: 20 h, Dose: 15 g/L (Zn^{2+}), 18 g/L (Cd^{2+}), pH: 6.0 (Zn^{2+}), 5.0 (Cd^{2+}), Initial conc.: 337 mg/L (Zn^{2+}), 634 mg/L (Cd^{2+})]	
CFA was collected from Duvha Thermal Power Plant, Mpumalanga, South Africa **(Class F)** **Properties:** S_{BET}: 1.19 cm^2/g Pore Volume: 0.001 cm^3/g Pore Size: NR CEC: 0.29 meq NH_4^+/g	**Zeolite A (ZA) and Sodalite (ZS)** Properties: S_{BET}: 29.5 cm^2/g (ZA), 7.81 cm^2/g (ZS) Pore Volume: 0.05 cm^3/g (ZA), 0.019 cm^3/g (ZS) Pore Size (nm): 6.71 nm (ZA), 8.96 nm (ZS) CEC: 3.1 meq NH_4^+/g (ZA), 0.98 meq NH_4^+/g (ZS) Application: NR	[37, 38]
CFA was from Thermal Power Plants in Bulgaria, TPP "AES Galabovo" **(Class F)** **Properties:** NR	**Nanozeolite Na-X (ZFH2)** Properties: S_{BET}: 486 cm^2/g Pore Volume: 0.31 cm^3/g Pore Size: 1.39-4.18 nm Yield: 89 wt.% Application: NR	[35]

NR: Not Reported.

Table 3.
Properties of zeolites derived from coal fly ash (CFA) and their applications.

LTA zeolite (10-MU-75-LTA) synthesized using combined microwave-ultrasonication, was found to exhibit 442 cm^2/g surface area. The ultrasonication enhances the dissolution of silica and alumina contents from the CFA into the alkaline gel mixture, which ultimately leads to the formation of the zeolites. In addition to that during the sonication, effects of acoustic cavitation along with the collapsing of micro-bubbles in the alkaline reaction mixture additionally enhance the rate of secondary nucleation, and quicken the mass transfer and expansion in the reactive surface area by means of fragmentation of solid crystals, which further help in reducing the time and temperature for the zeolite crystallization from CFA [77–79].

5. Conclusion and future perspectives

This review addresses the environmental problems caused due to fly ash generation and its mitigation measures *via* cost-effective and ecofriendly utilization of waste fly ash. Due to the presence of considerable amount of metal oxides, mainly SiO_2 and Al_2O_3 in fly ash, it is widely accepted as precursor of choice for the zeolite synthesis. The information provided in this review will be used as a guiding tool for the selection of fly ash source and an appropriate synthetic route to follow for the synthesis of a particular zeolite of desired properties for its targeted applications. A brief exposure to various synthetic approaches has been discussed in this chapter, which is currently in practice for zeolite synthesis. In recent times, the microwave and ultrasonication methods seem promising over conventional hydrothermal methods. However, still challenges exist for their successful large-scale implementation. Further studies and research efforts are required to enhance the feasibility to transfer the technology from lab scale to industrial scale. In general, various zeolites such as ZSM-5, Zeolite A, Zeolite Na-P, Zeolite X, Zeolite Y, Analcime are synthesized from fly ash. However, the synthesis of other zeolites having high surface area, pore volume, high acidity, and high cation exchange capacity and thus having commercial market values such as zeolite β, Mordenite, MCM-22 can be explored *via* systematic study of different synthetic parameters and methods. Moreover, there is a need to develop a process, which is associated with minimum waste generation and providing maximum utilization of the elemental composition of fly ash by extracting elements other than Si and Al, separately such as Fe_2O_3, TiO_2, trace amount of rare earth elements. Apart from these, emphases have been put in this chapter on fly ash utilization for generation of value-added materials such as zeolites and applications of fly ash-derived zeolites in the field of catalysis and environmental remediation. Future scope exists in executing large-scale application of fly ash in waste land reclamation, floriculture, and heavy metals recovery.

Author details

Henilkumar M. Lankapati[1], Kalpana C. Maheria[1*] and Ajay K. Dalai[2]

1 Department of Chemistry, Sardar Vallabhbhai National Institute of Technology, Surat, Gujarat, India

2 Department of Chemical and Biological Engineering, University of Saskatchewan, Saskatoon, SK, Canada

*Address all correspondence to: kcm@chem.svnit.ac.in

References

[1] Kotova OB, Shabalin IL, Shushkov DA, Kocheva LS. Hydrothermal synthesis of zeolites from coal fly ash. Advances in Applied Ceramics. 2016:**115**(3)152-157

[2] Belviso C, Cavalcante F, Javier Huertas F, Lettino A, Ragone P, Fiore S. The crystallisation of zeolite (X- and A-type) from fly ash at 25 °c in artificial sea water. Microporous Mesoporous Materials. 2012;**162**:115-121. DOI: 10.1016/j.micromeso.2012.06.028

[3] Kim K, Kim K, Asaoka S, Lee IC, Kim DS, Hayakawa S. Quantitative measurement on removal mechanisms of phosphate by class–F Fly ash. International Journal of Coal Preparation and Utilization. 2018;**00**(00):1-12

[4] Deng X, Qi L, Zhang Y. Experimental study on adsorption of hexavalent chromium with microwave-assisted alkali modified Fly ash. Water, Air, and Soil Pollution. 2018;**229**(1):18

[5] Appiah-Hagan E, Chen YW, Yu X, Arteca GA, Pizarro J, Mercier L, et al. Simple and energy-saving modifications of coal fly ash to remove simultaneously six toxic metal cations from mine effluents. Journal of Environmental Chemical Engineering. 2018;**6**: 5498-5509

[6] Karanac M, Đolić M, Veljović Đ, Rajaković-Ognjanović V, Veličković Z, Pavićević V, et al. The removal of Zn2+, Pb2+, and As(V) ions by lime activated fly ash and valorization of the exhausted adsorbent. Waste Management. 2018;**78** (November):366-378

[7] Chinh NT, Thai H, Tran TDM, Nguyen TT, Nguyen ON, Tran MT, et al. Adsorption of mercuric ion from aqueous solutions using modified Fly ash. Vietnam Journal of Science and Technology. 2018;**56**(6):688

[8] Lathiya DR, Bhatt DV, Maheria KC. Sulfated Fly-ash catalyzed biodiesel production from maize acid oil feedstock: A comparative study of Taguchi and box-Behnken design. ChemistrySelect. 2019;**4**(14):4392-4397

[9] bp Statistical Review 2022. bp Statistical Review of World Energy globally consistent data on world energy markets. The review is one of the most widely respected The Statistical of publications World Energy analyses and Review energy used from by the prior The Review academia , ha. 2022

[10] CEA Report on Fly Ash, New Delhi C. Report on fly ash generation at coal/lignite based thermal power stations and its utilization in the country for the year 2020–21. New Delhi; 2021. Available from: https://cea.nic.in/wp-content/uploads/tcd/2021/09/Report_Ash_Yearly_2020_21.pdf

[11] Senapati MR. Fly ash from thermal power plants—Waste management and overview. Current Science. 2011; **100**(12):1791-1794

[12] MoEF GOI. MoEF Gazette Notification dt. 1999. Available from: http://egazette.nic.in/WriteReadData/2020/221325.pdf

[13] MoEF GOI. MoEF Gazette Notification dt. 2003. Available from: https://egazette.nic.in/WriteReadData/2003/E_773_2011_007.pdf

[14] MoEF GOI. MoEF Gazette Notification dt. 2009. Available from: https://egazette.nic.in/WriteReadData/2009/E_1799_2011_002.pdf

[15] MoEFCC GOI. MoEFCC Gazette Notification dt2016. pp. 1-5. Available from: https://egazette.nic.in/WriteReadData/2016/167827.pdf

[16] MoEFCC GOI. MoEFCC Gazette Notification dt. Government of India; 2021. pp. 1-15. Available from: https://egazette.nic.in/WriteReadData/2021/226711.pdf

[17] Rastogi A, Paul VK. A critical review of the potential for fly ash utilisation in construction-specific applications in India. Environmental Research, Engineering and Management. 2020; **76**(2):65-75

[18] Lankapati HM, Lathiya DR, Choudhary L, Dalai AK, Maheria KC. Mordenite-type zeolite from waste coal Fly ash: Synthesis, characterization and its application as a sorbent in metal ions removal. ChemistrySelect. 2020;**5**(3): 1193-1198. DOI: 10.1002/slct.201903715

[19] Tumrani SH, Soomro RA, Zhang X, Bhutto DA, Bux N, Ji X. Coal fly ash driven zeolites for the adsorptive removal of the ceftazidime drug. RSC Advances. 2021;**11**(42):26110-26119

[20] Czuma N, Zarębska K, Motak M, Gálvez ME, Da Costa P. Ni/zeolite X derived from fly ash as catalysts for CO_2 methanation. Fuel [Internet]. 2020;**267** (May 2019):117139. Available from: https://www.sciencedirect.com/science/article/pii/S0016236120301344

[21] Missengue RNM, Losch P, Sedres G, Musyoka NM, Fatoba OO, Louis B, et al. Transformation of South African coal fly ash into ZSM-5 zeolite and its application as an MTO catalyst. Comptes Rendus Chimie. 2017;**20**(1):78-86

[22] Chen M, Zhu L, Dong Y, Li L, Liu J. Waste-to-resource strategy to fabricate highly porous whisker-structured Mullite ceramic membrane for simulated oil-in-water emulsion wastewater treatment. ACS Sustainable Chemistry & Engineering. 2016;**4**(4): 2098-2106

[23] Pavlović SM, Marinković DM, Kostić MD, Lončarević DR, Mojović L V, Stankovićć M V, et al. The chicken eggshell calcium oxide ultrasonically dispersed over lignite coal fly ash-based cancrinite zeolite support as a catalyst for biodiesel production. Fuel. 2021;**289** (July 2020):119912

[24] Pavlović SM, Marinković DM, Kostić MD, Janković-Častvan IM, Mojović LV, Stanković MV, et al. A CaO/zeolite-based catalyst obtained from waste chicken eggshell and coal fly ash for biodiesel production. Fuel. 2020; **267**(September):117171. DOI: 10.1016/j.fuel.2020.117171

[25] Vichaphund S, Aht-Ong D, Sricharoenchaikul V, Atong D. Characteristic of fly ash derived-zeolite and its catalytic performance for fast pyrolysis of Jatropha waste. Environmental Technology (United Kingdom). 2014;**35**(17):2254-2261

[26] Alzeer MIM, MacKenzie KJD. Synthesis and catalytic properties of new sustainable Aluminosilicate heterogeneous catalysts derived from Fly ash. ACS Sustainable Chemistry & Engineering. 2018;**6**(4):5273-5282

[27] ASTM C618-03. Standard Specification for Coal Fly Ash and Raw or Calcined Natural Pozzolan for Use West Conshohocken, PA. Annu B ASTM Stand. 2001;**2010**(C):3-6

[28] Yao ZT, Ji XS, Sarker PK, Tang JH, Ge LQ, Xia MS, et al. A comprehensive review on the applications of coal fly ash. Earth-Science Reviews. 2015;**141**:105-121

[29] Ti M, Gabrovska M. Miroslav Stanković.pdf.2020. pp. 1-24

[30] Surabhi. Fly ash in India : Generation Vis-à-Vis utilization and global perspective. International Journal of Applied Chemistry. 2017;**13**(1):29-52

[31] Yousuf A, Manzoor SO, Youssouf M, Malik ZA, Sajjad KK. Fly ash: Production and utilization in India-an overview. Journal of Materials and Environmental Science. 2020;**2020**(6):911-921

[32] Jangkorn S, Youngme S, Praipipat P. Comparative lead adsorptions in synthetic wastewater by synthesized zeolite A of recycled industrial wastes from sugar factory and power plant. Heliyon. 2022;**8**(4):e09323. DOI: 10.1016/j.heliyon.2022.e09323

[33] Das S, Yudhbir D. Chemistry and mineralogy of some indian fly ashes. Indian Concrete Journal. 2003;**77**: 1491-1494

[34] Sharma G, Mehla SK, Bhatnagar T, Bajaj A. Possible use of Fly ash in ceramic industries: An innovative method to reduce environmental pollution. International Journal of Modern Physics: Conference Series. 2013;**22**:99-102

[35] Bhattacharjee U, Kandpal TC. Potential of fly ash utilisation in India. Energy. 2002;**27**(2):151-166

[36] Dhadse S, Kumari P, Bhagia LJ. Fly ash characterization, utilization and government initiatives in India—A review. Journal of Scientific and Industrial Research (India). 2008;**67**(1):11-18

[37] Mintova S, Barrier N, Paillaud JL, Patarin J. Verified syntheses of zeolitic materials, initial materials for synthesis of zeolites: Introductory articles. In: Mintova S, Barrier N, editors. Synthesis Commission of the International Zeolite Association. 3rd ed. 2016. pp. 24-28. Available from: http://www.iza-online.org/synthesis/VS_3rdEd.pdf

[38] Chu Z, Liang J, Yang D, Chen H. Green chemical conversion of large-scale aluminosilicates into zeolites for environmental remediation under carbon-neutral pressure. Current Opinion in Green and Sustainable Chemistry. 2022;**36**:100632. DOI: 10.1016/j.cogsc.2022.100632

[39] Querol X, Alastuey A, López-Soler A, Plana F, Andrés JM, Juan R, et al. A Fast Method for Recycling Fly Ash: Microwave-Assisted Zeolite Synthesis. Environmental Science & Technology.. 1997;**31**(9):2527-2533

[40] Yang S, Yang L, Xiong P, Qian X, Nagasaka T. Ammonium continuous removal by zeolite P synthesized using fly ash combined with bacteria in aqueous solution. Environmental Science and Pollution Research. 2022. DOI: 10.1007/s11356-022-21350-3

[41] Belviso C, Cavalcante F, Fiore S. Synthesis of zeolite from Italian coal fly ash: Differences in crystallization temperature using seawater instead of distilled water. Waste Management. 2010;**30**(5):839-847. DOI: 10.1016/j.wasman.2009.11.015

[42] Ojha K, Pradhan NC, Samanta AN. Zeolite from fly ash: Synthesis and characterization. Bulletin of Materials Science. 2004;**27**(6):555-564

[43] Chang HL, Shih WH. A general method for the conversion of Fly ash into zeolites as ion exchangers for cesium. Industrial and Engineering Chemistry Research. 1998;**37**(1):71-78

[44] Yoshida A, Inoue K. Formation of faujasite-type zeolite from ground

Shirasu volcanic glass. Zeolites. 1986; **6**(6):467-473

[45] Amoni BC, Freitas ADL, Bessa RA, Oliveira CP, Bastos-Neto M, Azevedo DCS, et al. Effect of coal fly ash treatments on synthesis of high-quality zeolite A as a potential additive for warm mix asphalt. Materials Chemistry and Physics. 2022;**275**(May 2021):125197

[46] Kumar MM, Jena H. Direct single-step synthesis of phase pure zeolite Na–P1, hydroxy sodalite and analcime from coal fly ash and assessment of their Cs+ and Sr2+ removal efficiencies. Microporous Mesoporous Materials. 2022;**333**(January):111738. DOI: 10.1016/j.micromeso.2022.111738

[47] Bodart P, Nagy JB, Gabelica Z, Derouane EG. Factors governing the synthesis of zeolites from silicoaluminate hydrogels : A comparative study of the crystallization mechanisms of zeolites Y, mordenite and ZSM-5. Journal de Chimie Physique. 1986;**83**:777-790. DOI: 10.1051/jcp/1986830777

[48] Bo W, Hongzhu M. Factors affecting the synthesis of microsized NaY zeolite. Microporous and Mesoporous Materials. 1998;**25**(1–3):131-136

[49] Zhang Y, Zhou L, Chen L, Guo Y, Guo F, Wu J, et al. Synthesis of zeolite Na-P1 from coal fly ash produced by gasification and its application as adsorbent for removal of Cr(VI) from water. Frontiers of Chemical Science and Engineering. 2021;**15**(3):518-527

[50] Zhang X, Li C, Zheng S, Di Y, Sun Z. A review of the synthesis and application of zeolites from coal-based solid wastes. International Journal of Minerals, Metallurgy, and Materials. 2022;**29**(1):1-21

[51] Ma Z, Zhang X, Lu G, Guo Y, Song H, Cheng F. Hydrothermal synthesis of zeolitic material from circulating fluidized bed combustion fly ash for the highly efficient removal of lead from aqueous solution. Chinese Journal of Chemical Engineering. 2022; **47**:193-205. DOI: 10.1016/j.cjche.2021.05.043

[52] Shigemoto N, Hayashi H, Miyaura K. Selective formation of Na-X zeolite from coal fly ash by fusion with sodium hydroxide prior to hydrothermal reaction. Journal of Materials Science. 1993;**28**(17):4781-4786. DOI: 10.1007/BF00414272

[53] Murukutti MK, Jena H. Synthesis of nano-crystalline zeolite-a and zeolite-X from Indian coal fly ash, its characterization and performance evaluation for the removal of Cs+ and Sr2+ from simulated nuclear waste. Journal of Hazardous Materials. 2022; **423**(PA):127085. DOI: 10.1016/j.jhazmat.2021.127085

[54] Hui KS, Chao CYH. Pure, single phase, high crystalline, chamfered-edge zeolite 4A synthesized from coal fly ash for use as a builder in detergents. Journal of Hazardous Materials. 2006;**137**(1): 401-409

[55] Park M, Choi CL, Lim WT, Kim MC, Choi J, Heo NH. Molten-salt method for the synthesis of zeolitic materials II. Characterization of zeolitic materials. Microporous and Mesoporous Materials. 2000;**37**(1–2):91-98

[56] Chen W, Song G, Lin Y, Qiao J, Wu T, Yi X, et al. Synthesis and catalytic performance of Linde-type A zeolite (LTA) from coal fly ash utilizing microwave and ultrasound collaborative activation method. Catalysis Today. 2022;**397-399**(July 2021):407-418. DOI: 10.1016/j.cattod.2021.07.022

[57] Sireesha S, Agarwal A, Sopanrao KS, Sreedhar I, Anitha KL. Modified coal fly

ash as a low-cost, efficient, green, and stable adsorbent for heavy metal removal from aqueous solution. Biomass Convers Biorefinery. 2022. DOI: 10.1007/s13399-022-02695-8

[58] Makgabutlane B, Nthunya LN, Nxumalo EN, Musyoka NM, Mhlanga SD. Microwave irradiation-assisted synthesis of zeolites from coal Fly ash: An optimization study for a sustainable and efficient production process. ACS Omega. 2020;**5**(39): 25000-25008

[59] Plessis PW Du. Process design for the up-scale zeolite synthesis from South African coal fly ash. Cape Town: Cape Peninsula University of Technology; 2014. Available from: https://core.ac.uk/download/pdf/148364786.pdf

[60] Inada M, Tsujimoto H, Eguchi Y, Enomoto N, Hojo J. Microwave-assisted zeolite synthesis from coal fly ash in hydrothermal process. Fuel. 2005;**84**(12–13):1482-1486

[61] Behin J, Bukhari SS, Dehnavi V, Kazemian H, Rohani S. Using coal fly ash and wastewater for microwave synthesis of LTA zeolite. Chemical Engineering and Technology. 2014;**37**(9):1532-1540

[62] Belviso C, Cavalcante F, Fiore S. Ultrasonic waves induce rapid zeolite synthesis in a seawater solution. Ultrason Sonochem. 2013;**20**(1):32-36. DOI: 10.1016/j.ultsonch.2012.06.004

[63] Ju T, Jiang J, Meng Y, Yan F, Xu Y, Gao Y, et al. An investigation of the effect of ultrasonic waves on the efficiency of silicon extraction from coal fly ash. Ultrason Sonochem. 2020;**60** (May 2019):104765

[64] Ojumu TV, Du Plessis PW, Petrik LF. Synthesis of zeolite a from coal fly ash using ultrasonic treatment—A replacement for fusion step. Ultrason Sonochem. 2016;**31**:342-349. DOI: 10.1016/j.ultsonch.2016.01.016

[65] Maharana M, Sen S. Industrial solid waste based EU-12 Nanozeolite: Synthesis and characterisation. Waste and Biomass Valorization. 2022;**13**(3): 1695-1703. DOI: 10.1007/s12649-021-01553-4

[66] Itabashi K, Kamimura Y, Iyoki K, Shimojima A, Okubo T. A working hypothesis for broadening framework types of zeolites in seed-assisted synthesis without organic structure-directing agent. Journal of the American Chemical Society. 2012;**134**(28):11542-11549

[67] Lobo RF, Davis ME. CIT-1: A new molecular sieve with intersecting pores bounded by 10- and 12-rings. Journal of the American Chemical Society. 1995; **117**(13):3766-3779

[68] Cundy CS, Cox PA. The hydrothermal synthesis of zeolites: Precursors, intermediates and reaction mechanism. Microporous and Mesoporous Materials. 2005;**82**(1–2):1-78

[69] Thompson RW, Dyer A. Nucleation of zeolite NaA crystals in hydrothermal systems. Zeolites. 1985;**5**(5):302-308

[70] Kacirek H, Lechert H. Growth of the zeolite type NaY. Journal of Physical Chemistry. 1975;**79**(15):1589-1593. DOI: 10.1021/j100582a024

[71] Liu Z, Wakihara T, Oshima K, Nishioka D, Hotta Y, Elangovan SP, et al. Widening synthesis bottlenecks: Realization of ultrafast and continuous-flow synthesis of high-silica zeolite SSZ-13 for NOx removal. Angewandte Chemie. 2015;**54**(19):5683-5687

[72] Liu Z, Wakihara T, Nishioka D, Oshima K, Takewaki T, Okubo T.

One-minute synthesis of crystalline microporous aluminophosphate (AlPO4-5) by combining fast heating with a seed-assisted method. Chemical Communications. 2014;**50**(19): 2526-2528

[73] Zhang C, Zhang R, Cao Y. Value-added utilizations of fly ash via the extraction of rare earth elements and silica as well as the synthesis of zeolites. SSRN. 2022. Available from: https://papers.ssrn.com/sol3/papers.cfm?abstract_id=4077853

[74] Makgabutlane B, Nthunya LN, Musyoka N, Dladla BS, Nxumalo EN, Mhlanga SD. Microwave-assisted synthesis of coal fly ash-based zeolites for removal of ammonium from urine. RSC Advances. 2020;**10**(4): 2416-2427

[75] Budiman AW, Nugrahaningtyas KD, Purnama SDPW, Nabila AK, Gerard PI, Wulansari DWT, et al. The effect of alumina addition on the synthesis of zeolite from fly ash by hydrothermal method. Journal of Physics Conference Series. 2022;**2190**(1):012003

[76] Niwa M, Katada N, Okumura K. Characterization and Design of Zeolite Catalysts. Berlin, Heidelberg: Springer Berlin Heidelberg; 2010. pp. 1-181. (Springer Series in Materials Science; vol. 141)

[77] Boycheva S, Marinov I, Miteva S, Zgureva D. Conversion of coal fly ash into nanozeolite Na-X by applying ultrasound assisted hydrothermal and fusion-hydrothermal alkaline activation. Sustainable Chemistry and Pharmacy. 2020;**15**(August 2019):100217. DOI: 10.1016/j.scp.2020.100217

[78] Andaç Ö, Tatlier M, Sirkecioğlu A, Ece I, Erdem-Şenatalar A. Effects of ultrasound on zeolite a synthesis. Microporous and Mesoporous Materials. 2005;**79**(1–3):225-233

[79] Belviso C, Cavalcante F, Lettino A, Fiore S. Effects of ultrasonic treatment on zeolite synthesized from coal fly ash. Ultrason Sonochem. 2011;**18**(2):661-668. DOI: 10.1016/j.ultsonch.2010.08.011